Daniel Müller

RF Probe-Induced On-Wafer Measurement Errors in the Millimeter-Wave Frequency Range

Karlsruher Forschungsberichte
aus dem Institut für Hochfrequenztechnik und Elektronik

Herausgeber: Prof. Dr.-Ing. Thomas Zwick

Band 89

Eine Übersicht aller bisher in dieser Schriftenreihe erschienenen Bände finden Sie am Ende des Buches.

RF Probe-Induced On-Wafer Measurement Errors in the Millimeter-Wave Frequency Range

by
Daniel Müller

KIT Scientific Publishing

Karlsruher Institut für Technologie
Institut für Hochfrequenztechnik und Elektronik

RF Probe-Induced On-Wafer Measurement Errors in the Millimeter-Wave Frequency Range

Zur Erlangung des akademischen Grades eines Doktor-Ingenieurs von der KIT-Fakultät für Elektrotechnik und Informationstechnik des Karlsruher Instituts für Technologie (KIT) genehmigte Dissertation

von Dipl.-Ing. Daniel Müller

Tag der mündlichen Prüfung: 06. Juni 2018
Referent: Prof. Dr.-Ing. Ingmar Kallfass
Korreferent: Prof. Dr.-Ing. Thomas Zwick

Impressum

KIT Scientific Publishing

Karlsruher Institut für Technologie (KIT)
KIT Scientific Publishing
Straße am Forum 2
D-76131 Karlsruhe

KIT Scientific Publishing is a registered trademark of Karlsruhe Institute of Technology.
Reprint using the book cover is not allowed.

www.ksp.kit.edu

Print on Demand 2018 – Gedruckt auf FSC-zertifiziertem Papier

ISSN 1868-4696
ISBN 978-3-7315-0822-9
DOI 10.5445/KSP/1000084392

Editor's Preface

Due to the rapid technological advances of semiconductor technologies (SiGe, CMOS, GaAs, GaN, InP, etc.) in recent years, the millimeter-wave frequency region gained enormous interest in academia as well as in industry. Along with novel and economic packaging technologies, millimeter-wave applications – such as automotive radar, industrial sensors, high-speed communication and medical technology – reach for the mass market. However, not only the realization of the modules but also the corresponding metrology remains challenging at these frequencies. The usability of classical connectors above 100 GHz is not only expensive but is also a limiting factor in terms of size, bandwidth and measurement accuracy. These limitations led to the development and widespread use of RF probes for characterizing the performance of integrated circuits (ICs). Yet, in contrast to the commonly used coaxial or waveguide transmission lines, such RF probes do not have a single mode of propagation at the interface to the chip, which causes an increase in measurement errors towards higher frequencies.

This is exactly where the work of Dr. Daniel Müller comes into play. In the framework of his activities at the Institute of Radio Frequency Engineering and Electronics (IHE) at the Karlsruhe Institute of Technology (KIT), he started with the design of integrated phase shifters for frequencies above 100 GHz. He soon realized that the existing metrology is limiting the accuracy of the measurement results and thus decided to devote himself to the improvement of this measurement methodology. For the first time Dr. Müller investigated the systematic measurement errors to the device under test caused by the interaction between the RF probe's geometry and the chosen measurement setup, which is many times falsely attributed to process and fabrication tolerances. By creating 3D models of the RF probes suitable for electromagnetic simulations, he was able to replicate the occurring effects at the interface and therefore achieve a much better agreement between the simulation and the measured behavior. Based on this method Dr. Müller was able to identify the origin of the measurement errors and he developed pad structures, which offer a drastically improved performance up to 330 GHz. Furthermore, using an adapted 2nd-tier calibration procedure he successfully corrected measurement errors caused by crosstalk of the RF probes. The work of Dr. Daniel Müller thus represents an important innovation to the state of the art. I am sure that his novel measurement methods will draw much attention and find many users worldwide. For Dr. Müller, with his creativity and great organizational skills, I wish him further much success in his scientific engineering career and economic endeavors.

Prof. Dr.-Ing. Thomas Zwick
- Institute Director -

RF Probe-Induced On-Wafer Measurement Errors in the Millimeter-Wave Frequency Range

Zur Erlangung des akademischen Grades
Doktor der Ingenieurwissenschaften
der Fakultät für Elektrotechnik und Informationstechnik
Karlsruher Institut für Technologie (KIT)

genehmigte
Dissertation
von

Dipl.-Ing. Daniel Müller

Tag der mündlichen Prüfung:	06. Juni 2018
Referent:	Prof. Dr.-Ing. Ingmar Kallfass
Korreferent:	Prof. Dr.-Ing. Thomas Zwick

Kurzfassung

In den vergangenen Jahren ermöglichte der rasante technologische Fortschritt von monolithisch integrierten Schaltungen (MMICs) die Erschließung von Frequenzen bis in den Terahertz (THz) Bereich.
Dieser Fortschritt wurde maßgeblich durch die starke Skalierung der integrierten Transistoren erreicht, welche zu geringen parasitären Effekten und hohen Transitfrequenzen führt. Die notwendige Charakterisierung der MMICs wird hauptsächlich mit Hochfrequenz-Messspitzen, sogenannten RF Probes durchgeführt. Jedoch können weder die verwendeten RF Probes noch die dazugehörigen Kontaktpads angemessen skaliert werden, um im Millimeterwellen-Frequenzbereich (30 - 300 GHz) und darüber ein transparentes Verhalten, ohne Beeinflussung der MMIC Performanz, zu gewährleisten. Durch die in Bezug auf die Wellenlänge und im Vergleich zu den restlichen Schaltungskomponenten großen Dimensionen der RF Probes und Kontaktpads, sind diese anfällig für parasitäre Effekte wie höhere Moden, Abstrahlung und Substrateinkopplung, welche die Messergebnisse und daraus abgeleitete Modelle verfälschen. Die Auswirkungen der parasitären Effekte werden oft fälschlicherweise Fabrikations- und Prozesstoleranzen zugeschrieben. Damit zwischen dem Einfluss der RF Probes und eventuellen Fabrikations- und Prozesstoleranzen unterschieden werden kann, werden in dieser Arbeit Modelle der RF Probes für die elektromagnetische Feldsimulation erstellt. Dies erfordert als Grundlage exakte Modelle der zu untersuchenden MMICs, deren Erstellung für die verwendete Indium-Galium-Arsenid (InGaAs) Technologie ebenfalls vorgestellt wird. Der interne Aufbau der erstellten RF Probe Modelle und die

Vorgehensweise zum De-Embedden der RF Probe Modelle aus den simulierten Ergebnissen werden erörtert. Die entwickelten Modelle werden unter anderem dazu verwendet, die Anpassung und Verluste von Kontaktpads sowie die Genauigkeit von Kalibrierstandards zu verbessern. Weiterhin wird das Übersprechen zwischen RF Ein- und Ausgangsprobe untersucht, sowie der Einfluss auf die Messergebnisse analytisch abgeschätzt. Abschließend wird ein neuartiges Verfahren vorgestellt, um die durch Übersprechen der RF Probes verfälschten Messergebnisse zu korrigieren, welches erfolgreich im Frequenzbereich von 200 bis 330 GHz auf einen integrierten diskreten Phasenschieber und einen integrierten einpoligen Wechselschalter angewendet wird.

Abstract

In recent years monolithic microwave integrated circuits (MMICs) have increased their performance drastically, approaching operational circuits in the terahertz region. This advancement is mainly based on the shrinking of integrated transistors, which minimizes parasitic effects and increases maximum transit and oscillation frequencies. To characterize such MMICs it is common to use radio frequency (RF) probe based measurement systems. However, compared to the integrated transistors and circuit elements, the RF probes and corresponding contact pads are large in terms of the corresponding wavelength, which makes these structures prone to parasitic effects such as radiation, unwanted modes or substrate coupling. These parasitic effects distort the measurement results and are often falsely attributed from fabrication and process tolerances. To clearly distinguish the effect of the RF probes to fabrication and process tolerances this work presents a novel method, which is based on accurate models for electromagnetic (EM) field simulations of the same RF probes used in the measurement setup. This technique requires the availability of an accurate EM model of the device under test (DUT), which is discussed and realized for the utilized indium gallium arsenide (InGaAs) technology. The development of the different RF probe models, used for signal excitation, is presented and the corresponding procedure of de-embedding the RF probes out of the simulation results is discussed in detail. The novel technique is applied to the successful optimization of contact pads in terms of bandwidth and loss as well as calibration structures for high accuracy measurements. Furthermore, the crosstalk between input and output RF probe in different measurement environments is investigated and the influence on measurements is estimated analytically. Finally, a technique for correction of crosstalk distorted measurement results is presented and successfully verified in the frequency range from 200 to 330 GHz for an integrated discrete phase shifter and an integrated single pole double throw (SPDT) switch.

Acknowledgement

First of all I would like to thank Prof. Dr.-Ing. Ingmar Kallfass and Prof. Dr.-Ing. Thomas Zwick for supervising this work while giving me the scientific freedom and encouragement to follow my own ideas. I am personally indebted to the staff of the Fraunhofer IAF circuit design and technology department for many fruitful discussions and manufacturing of the integrated circuits. My grateful acknowledgement goes to Dr.-Ing. Ulrich Lewark and Jochen Schäfer for proofreading of the manuscript and moreover for the constructive criticism and discussions over the past years. Furthermore, I am grateful to Katrijn Dill-Engelen who amiably proof read the manuscript for language, so that I am now confident enough to publish this work in English. Generally, it is a pleasure to thank all colleagues at IHE for the supportive and friendly atmosphere throughout the whole time. I would also like to thank Jerzy Kowalewski and my office mate Akanksha Bhutani for uncountable refreshing technical and non-technical conversations. Special thanks goes to the technical staff of the IHE, namely Andreas Lipp, Ronald Vester, Andreas Gallego and Mirko Nonnenmacher for untiringly realizing my sometimes time-consuming ideas. Furthermore, I would like to thank Angela Ziemba for managing the bureaucratic jungle of project coordination, business trips and daily business. My very special thanks goes to my parents and my sister, who supported me unconditionally over the past three decades. Above all, my deepest gratitude goes to Stefanie Remer for being my daily source of inspiration and persistently reminding me that there is a world besides engineering.

Contents

List of abbreviations and variables

ADS	Advanced Design System
APD	active power divider
BCB	benzo-cyclo-butene
CAD	computer-aided design
CDR	conductor, dielectric and radiation
CG	common gate
CM	Cascade Microtech Inc.
CPW	coplanar waveguide
CS	common source
CST	Computer Simulation Technology AG
CMP	chemical mechanical polishing
DC	direct current
DGS	defected ground structure
DPS	digital phase shifter
DUT	device under test
EM	electromagnetic
FDTD	finite-difference time-domain
FOM	figure of merit
GaAs	Gallium Arsenide
GGB	GGB Industries Inc.
IAF	Institute for Applied Solid State Physics

InGaAs indium gallium arsenide

InP indium phosphide

ISS impedance standard substrate

LRRM Line-Reflect-Reflect-Match

mHEMT metamorphic high electron mobility transistor

MIM metal-insulator-metal

mmW millimetre wave

MMIC monolithic microwave integrated circuit

mTRL multiline TRL

MWS Microwave Studio®

NiCr nickel-chromium

OML Oleson Microwave Labs Inc.

PCB printed circuit board

PI polyimide

PTFE polytetrafluoroethylene

PDK process design kit

RF radio frequency

RTPS reflective-type phase shifter

SEM scanning electron microscope

SiN silicon nitride

SOLT Short-Open-Load-Thru

SPDT single pole double throw

TFMS thin film microstrip

TRL Thru-Reflect-Line

VGA variable gain amplifier

VNA vector network analyzer

WG waveguide

1 Introduction

1.1 Motivation

The classical way of characterizing microwave and millimetre wave (mmW) components is to embed the device under test (DUT) into a rectangular waveguide (WG) and perform the characterization in a WG based measurement system. This has the advantage that, within the specified frequency range of the corresponding waveguide band, there is only a single mode of propagation present at the WG interface and there is no ambiguity of what is being measured. After the successful characterization, these individual devices can be connected to more complex systems. However, this technique was highly time and money consuming, which led to the development of radio frequency (RF) probes, which act as an adapter between the WG and the pad on the DUT. Therefore, using RF probes allows to directly measure the DUT performance without the need of prior packaging. Nowadays almost all measurements of monolithic microwave integrated circuits (MMICs) are performed by the use of such RF probes. A typical RF probe based measurement is shown in Fig. 1.1a. A close-up photograph of a common RF probe tip is shown in Fig. 1.1b.

The actual probe tips of such RF probes usually consist of three signal conductors in a ground-signal-ground configuration, which create a mechanical and electrical contact to the DUT. This is well proven at low mmW frequencies, where the wavelength is much larger than the dimensions of the RF probe. Approaching the high mmW frequency range on the other hand, a number of measurement and calibration issues arise. These issues are, besides fabrication tolerances of the RF probe itself, a result of the decreasing wavelengths, which eventually end up in the same order of magnitude as the RF probe tip and contact pad dimensions. Compared to waveguide based calibrations and measurements, where a single mode of propagation is present at the interface of two devices, RF probe tip to pad interconnects are, due to the relatively large dimensions, prone to multi mode propagation and parasitic effects. This issue is sketched in Fig. 1.2.

Figure 1.1: Photograph of a (a) typical on-wafer RF probe based measurement and (b) close-up of a common RF probe tip.

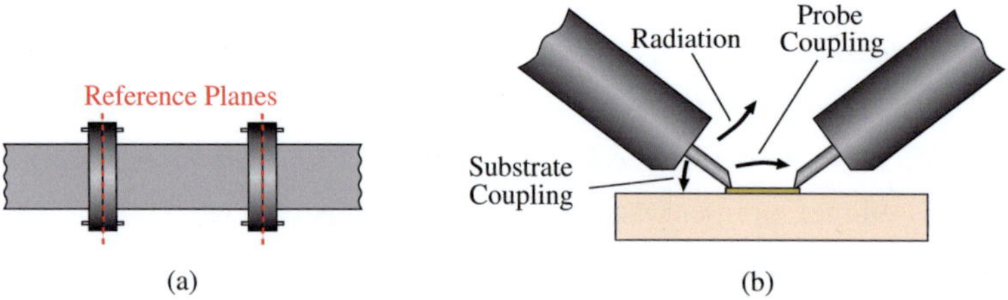

Figure 1.2: Exemplary measurement of a simple line section for (a) WG based measurements and (b) RF probe based measurements.

Effects like substrate coupling, radiation and RF probe coupling are inevitable and do not only limit the accuracy of the calibration but also have an impact on the overall circuit behaviour. In addition to these systematic errors, further deterioration such as poor RF probe contact, bad planarization, inaccurate RF probe placement, etc. may occur if an untrained operator is carrying out the measurements. While these effects are related to RF probe based measurements, there are also several parasitic effects of the DUT itself, which arise at mmW frequencies and worsen the circuit performance. Generally, components can be modelled using lumped elements as long as their spatial extend is smaller than approximately one tenth of a wavelength [Kai05]. However, going up in frequency this requirement cannot be fulfilled any longer. For example at 300 GHz one tenth of the free space wavelength is equal to 100 µm but for on-chip integrated transmission lines, where part of the electric field propagates inside the substrate, the effective wavelength shortens by the velocity factor $1/\sqrt{\varepsilon_{r,\text{eff}}}$. $\varepsilon_{r,\text{eff}}$ is the effective dielectric constant and can be approximated for a coplanar waveguide (CPW) transmission line by [Sim01]

$$\varepsilon_{r,\text{eff}} = \frac{\varepsilon_r + 1}{2} \tag{1.1}$$

with ε_r being the dielectric constant of the used substrate. For a CPW line on Gallium Arsenide (GaAs) substrate this results in an effective permittivity $\varepsilon_{r,\text{eff}}$ of approximately 6.95, resulting in a guided wavelength of 380 µm. Therefore, for the design of MMICs operating in the high mmW frequency range it is necessary to use distributed circuit element models. Furthermore, the complex layer stack of MMIC technologies, consisting of several metal and dielectric layers, degrades the accuracy of analytical models e.g. for calculating the characteristic impedance of a transmission line. Other issues when going to mmW frequencies are parasitic modes, which start to propagate inside the substrate [God92]. One way to shift the cutoff frequencies of these parasitic modes is to decrease the thickness of the substrate. However, very thin wafers are mechanically fragile and tend to crack easily. Another way of substrate mode suppression is to add through-substrate vias with a close spacing between each other, depending on the frequency. All these effects make it inevitable at mmW frequencies to verify the electrical characteristics and robustness of a circuit design by full wave electromagnetic (EM) simulations, which are in combination with precise RF probe tip models, capable of precisely predicting the actual circuit behaviour.

1.2 On-wafer measurements

Measurements at mmW frequencies are predominantly performed using RF probes in conjunction with manual or automatic probing stations for positioning and contacting the DUTs. The RF probes realize a mechanical and electrical contact between probe and DUT to introduce the measurement signal and if necessary provide bias to the circuit. In contrast to non-contact measurement systems, which require very specific pad structures [Jam+14; ZG11], the pads of RF probe based measurements can be designed to be not only suitable for measurements, but also for wire bonding or flip chip packaging. This is essential for high accuracy RF application, since fabrication tolerances require characterization of each die before packaging. Due to this advantage, contact based RF probes were developed rapidly in recent years and are nowadays available up to 1.1 THz with a contact pin pitch down to 25 µm [Bau+14]. Vector network analysers (VNAs), which generate and measure the test signals, are currently limited to frequencies below 70 GHz and therefore so called VNA extension modules are used to extend the frequency range into higher bands. These modules, based on frequency multipliers, subharmonic mixers and couplers are available up to 1.5 THz nowadays [VDI].

1.2.1 Used on-wafer measurement system

All measurements presented in this work were performed on the measurement system described in the following paragraph, which operates up to a maximum frequency of 330 GHz. A Keysight PNA-X N5247A VNA is used to record the measurements, internally covering the frequency range from 10 MHz to 67 GHz. To address the WR3 frequency range (220 to 325 GHz), extension modules manufactured by Oleson Microwave Labs Inc. (OML) [OML] were connected to the VNA using a Keysight N5262A mmW controller. In addition to the specified WR3 frequency range, the used WR3 extension modules generate sufficient output power from 200 to 330 GHz and can be calibrated with only little decrease in dynamic range at the unspecified frequencies. The measurement setup is shown in Fig. 1.3.

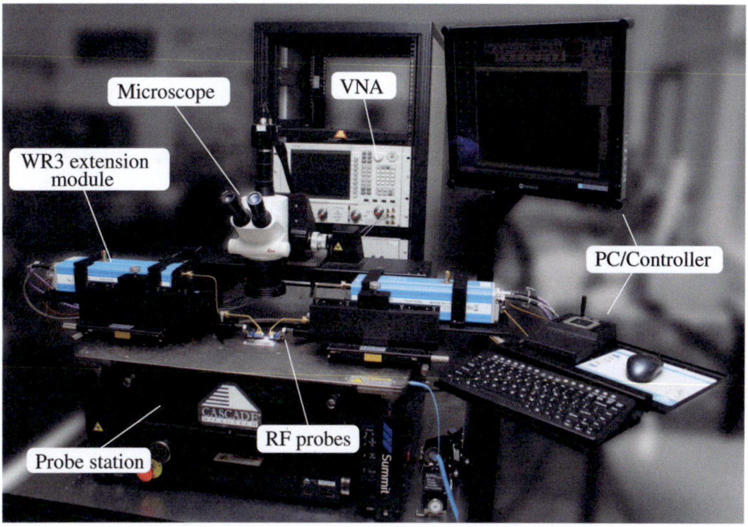

Figure 1.3: Photograph of the used on-wafer measurement setup.

To contact the DUTs, two different RF probes were used: Cascade Microtech Inc. (CM) Infinity Probes® with a pitch of 100 µm and GGB Industries Inc. (GGB) Picoprobes® with a pitch of 80 µm. Both RF probes are specified for the investigated frequency range but differ in their physical realization, see Fig. 1.4. Detailed descriptions of the used probes are discussed along with the corresponding EM models in section 3.1.

For a better readability the Cascade Microtech Inc. Infinity Probes® are referred to in the following work as CM probes while the GGB Industries Inc. Picoprobes® are referred to as GGB probes.

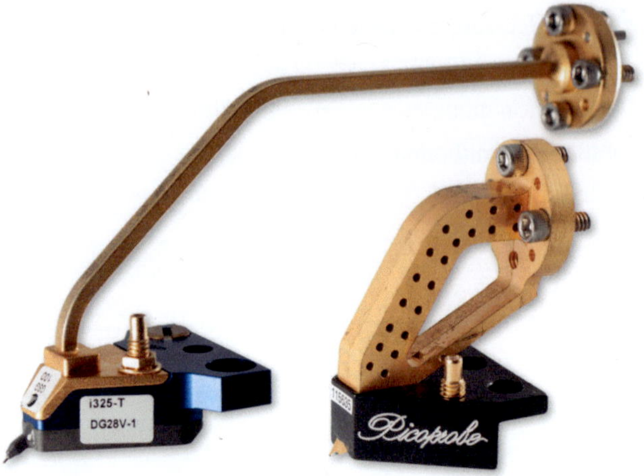

(a) (b)

Figure 1.4: Pictures of the used RF probes. (a) Cascade Microtech Inc. Infinity Probe® (b) GGB Industries Inc. Picoprobe®.

The probes are mounted on positioners which allow fine adjusting of the probe position in X, Y and Z direction with an accuracy down to 3 µm. Furthermore these positioners allow to adjust the tilt of the RF probes, which is essential to align the probe to the DUT and ensure a proper contact of all three contact pins. Since measuring the circuits on the wafer before separation may suppress parasitic effects, due to the extended substrate, in this work only individual dies were measured. This is also a better representation of the environment for packaged MMICs. Furthermore, this restricts the EM simulation volume and eliminates any possible influence from adjacent structures. Since the probes have a mechanical contact to the DUT it is necessary to fix the MMIC in order to prevent shifting or flipping of the die. While whole wafers or large dies can be fixed applying vacuum to the probe station chuck, this is not possible for small dies since the diameter of the vacuum holes is in the same order as the chip size (in this work down to $500 \times 500\,\mu m^2$). Usually, such small dies are attached to a metal plate using an electrically conductive adhesive, which can then be fixed by vacuum. However, during this process it has to be ensured that the adhesive is strictly restricted to the area under the MMIC as a thin layer and no overflow occurs at the edges of the MMIC. Any overflow may influence the RF characteristics of the MMIC in an unknown manner. This strict requirement could not be guaranteed with the given laboratory equipment and therefore to measure the small dies they were first fixed with direct current (DC) probes, usually used for bias supply, at the upper and lower edge of the MMIC and then contacted with the RF probes. For passive circuits without DC bias, the DC probes were lifted up prior to the measurement to not alter the measured circuit performance. With this technique the MMICs are directly placed on the measurement chuck, which simplifies EM re-simulation of the measurement environment drastically.

1.2.2 Calibration of the measurement system

In order to separate the influence of the measurement system from the DUT behaviour, a calibration has to be performed each time a specific measurement configuration is set up. Sometimes this task is also called error correction. The basic calibration concept is shown in Fig. 1.5 and is based on two error two-ports G and H, which contain the system response of test port 1 and test port 2, respectively. The VNA measurement values equal a_{G2}, b_{G2}, a_{H2}, b_{H2}, which in terms of transfer matrix parameters is given by:

$$\mathbf{T}_{\text{meas}} = \mathbf{T}_{\text{G}} \cdot \mathbf{T}_{\text{DUT}} \cdot \mathbf{T}_{\text{H}} \tag{1.2}$$

with $\mathbf{T}_{\text{G}}$ and $\mathbf{T}_{\text{H}}$ being the transfer matrix of the two error ports and $\mathbf{T}_{\text{DUT}}$ representing the transfer matrix of the DUT. The purpose of the calibration algorithm is first to determine the matrices $\mathbf{T}_G$ and $\mathbf{T}_H$. After this the actual DUT response $\mathbf{T}_{DUT}$ can then be extracted with inverse matrix calculations:

$$\mathbf{T}_{DUT} = \mathbf{T}_G^{-1} \cdot \mathbf{T}_{meas} \cdot \mathbf{T}_H^{-1} \tag{1.3}$$

Common implementations of calibration algorithms are Short-Open-Load-Thru (SOLT), Line-Reflect-Reflect-Match (LRRM) and Thru-Reflect-Line (TRL). Depending on the calibration algorithm, the error terms are calculated by measuring a set of known or partially known standards. For on-wafer RF probe measurements these calibration standards are embedded on an impedance standard substrate (ISS) whereas depending on the frequency range and probe pitch the probe manufacturer provides a specific ISS fitting the requirements. Furthermore, for RF probe based measurements at mmW frequencies the calibration algorithm has to be selected carefully [Hie07], as described below.

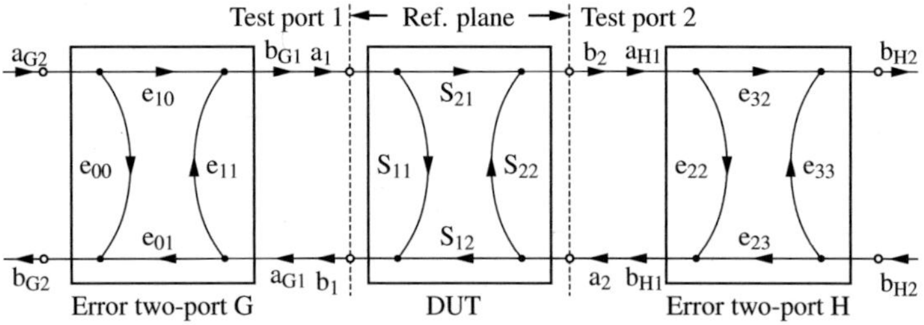

Figure 1.5: 7-term calibration error model [Hie07].

Short-Open-Load-Thru

The first published solution for determining the error terms was the SOLT method [KS71], which is based on measuring known reflect standards (open, short and match) as well as a direct connection (thru) between both ports. However, using this technique for accurate calibrations it is essential to either provide ideal standards or to have fully known standards. Both of these requirements are very challenging at mmW frequencies and can usually not be fulfilled. Furthermore, this calibration technique is sensitive to the accuracy of the RF probe placement [SH02]. Therefore, SOLT calibrations are only used at low mmW frequencies.

Line-Reflect-Reflect-Match

The LRRM method was specifically developed for use in on-wafer measurement scenarios [DJS90]. The novelty of the algorithm is the determination of the series inductance for the match standard during the calibration process which increases the calibration accuracy. Another advantage, especially for on-wafer measurements, is that the technique does not require lines of different length and therefore the distance between the RF probes is fixed for all calibration standards, which increases repeatability. The disadvantage is that the load standard is only measured at one of the VNA ports, which may lead to calibration errors at the second port.

Thru-Reflect-Line

The development of the TRL technique allowed the use of standards which were neither ideal nor fully known [EH79]. Three standards, two lines of different length and a reflect (open or short) need to be measured in order to calculate the error terms. Since the characteristic impedance of the line defines the reference impedance of the calibrated measurements, it has to be well defined and close to the system impedance, which is usually 50 Ω. The phase of the reflect standard has to be known to at least $\pm 90°$, i.e. depending on the location of the reflect position an approximate length offset must be given. These relaxed requirements allow easy integration of own calibration standards on the utilized MMIC technology. It has to be noted that, due to the different length of the calibration lines, the probe distance has to be changed during the calibration procedure, which if handled improperly may distort the calibration accuracy. Another disadvantage of the TRL technique

is the frequency band limitation which requires a phase difference of the thru and line standard between 20° and 160°. However, in waveguide based mmW measurements the usable frequency range is limited by the cutoff frequency of the lowest order and the next higher order mode of the waveguide. For standard rectangular waveguides this results in a usable frequency ratio of approximately 2 whereas the TRL calibration operates over a frequency ratio of approximately 8. If it is necessary to overcome the TRL bandwidth limitation, an adapted multiline TRL (mTRL) method – making use of multiple lines with different lengths – was presented in [Mar91].
In this work the TRL calibration was selected for all measurements and simulations, due to the possibility to easily use own calibration standards.

1.3 State of the art

To accurately predict the behaviour of high frequency circuits and components, EM simulations have a long tradition in RF and mmW engineering. The foundation for modern EM simulations was laid by [Yee66] who introduced the basic finite-difference time-domain (FDTD) algorithm. Since then, EM simulations have evolved rapidly and many commercial simulators are available, making EM analysis of circuit components daily routine to RF engineers.
For the investigations presented in this work the commercially available software suite Computer Simulation Technology AG (CST) Microwave Studio® [Tec17] was chosen, but others may be adopted as well. To simulate the characteristics of the DUT, the incident signals must be excited. The recommendation of CST for the excitation of CPW structures is the use of so called WG ports, which are exemplary shown in Fig. 1.6a. These ports first calculate the possible WG port modes using a 2D eigenmode solver, which are then used to excite the circuit. Even though this is a robust method for exciting single mode structures, this cannot be applied for the RF probe to contact pad interconnect, where a single mode of propagation

cannot be assumed. Therefore, to improve the simulation accuracy and investigate various effects, which arise from the interconnect, methods for more realistic signal excitation of the RF probe to pad interconnect were investigated in recent years. The work presented in [Sch+11] is proposing a novel approach where the signal is excited using a bridge of perfect conducting material between the ground planes and a discrete port to excite the inner conductor. This is exemplarily shown in Fig. 1.6b.

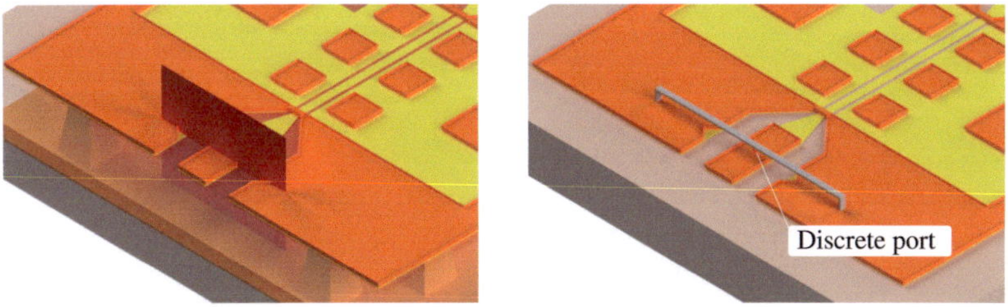

(a) (b)

Figure 1.6: (a) WG port excitation (transparent substrate for clarity) and (b) bridge + discrete port excitation.

They show a better prediction of the loss in CPW lines up to 110 GHz. This approach is picked up by [SGA13] and extended into the WR3 frequency range (220 - 325 GHz) but is only used to excite simple uniform coplanar transmission line structures. To investigate both excitation techniques, the WG port and the bridge with a discrete port, a simple test structure consisting of a transmission line and two contact pads was measured and simulated in the frequency range from 10 MHz to 330 GHz. The results are summarized in Figs. 1.7a and 1.7b for the magnitude of S_{11} and S_{21}, respectively. Two different RF probes were used for the measurements to verify the results. Comparing the curves of the input reflection coefficient S_{11} shows that the bridge method approximately describes the measured behaviour up to 170 GHz while the WG port results are starting to deviate at approximately 70 GHz. Above 170 GHz a strong deviation between measurement and simulation can be observed for both excitation methods. For the transmission coefficient S_{21}, a drop in magnitude around 300 GHz is predicted by the bridge excitation method but not if WG ports are used for signal excitation. However, this drop is verified by the measurements with both RF probes.

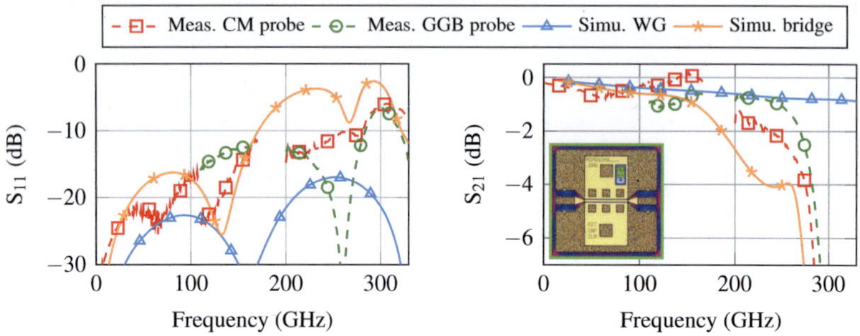

(a) $|S_{11}|$ (b) $|S_{21}|$

Figure 1.7: Comparison of measurement and simulation results of a transmission line test structure using different excitation methods.

While it seems to be possible by using the bridge method to simulate the coarse influence of multimode propagation and other parasitic effects on the return loss and insertion loss, detailed results cannot be achieved by either simulation method. Furthermore, the measurements indicate that the choice of RF probe has an influence on the measurement results, which is especially visible in the deviating reflection coefficient S_{11}. It is therefore necessary to include a model of the RF probe tips into the simulation domain, to comprehend the interaction between contact pad and RF probe. There are several publications which consider the influence of the RF probe by means of EM simulations. To predict the impact of the RF probes, [GS15] investigated the influence of different substrates between calibration and measurement up to 110 GHz by means of 3D EM simulation, using a simplified RF probe model. They successfully minimized the residual error by up to 50 % for a calibration transfer from alumina ($\varepsilon_r = 9.6$) to the DUT with $\varepsilon_r = 4.1$.

A more realistic RF probe model was used by [Wil+14a] where the authors evaluated the impact of probe crosstalk to the DUT and proposed a correction method for frequencies up to 110 GHz. Another approach was presented in [Saf05] in which the whole RF probe from connector to RF probe tip was 3D EM simulated and characterized. The authors showed a good agreement to measurements up to frequencies of 30 GHz with special focus of the unwanted microstrip mode between the RF probe outer conductor and the underlying metal chuck.

In conclusion, the embedding of RF probe models into EM simulations seems to improve the simulation results accuracy. However, the available publications on this topic are limited to frequencies below 110 GHz and only coarse models of the used RF probes were created for the EM simulations. The goal of this work is to increase the frequency range up to 330 GHz and to develop very accurate models of the used RF probes. This allows to precisely analyse the RF probe influence to the DUT and therefore to reveal the true circuit performance.

1.4 Goals and organization of the work

The goal of this work is to investigate parasitic effects in RF probe based on-wafer measurements above 200 GHz, which are observed frequently but cannot be comprehended nowadays with the existing simulation methods. These unwanted effects lead to a deviation from simulation to measurement, which usually results in a deterioration of the DUT RF performance but may also lead to complete failure. To have full control of the environment these investigations are carried out by 3D EM simulations. The pursued approach, to improve the agreement between measurement and simulation, is to embed RF probe tip models into the simulation domain and therefore replicate the real world environment as closely as possible. The present work is organized as follows:

Chapter 2 describes the utilized InGaAs MMIC technology of the Fraunhofer Institute for Applied Solid State Physics (IAF) and the challenges of non-planarized wafers for EM simulation purposes are discussed. After the features and layer stack of the used technology are discussed the technique for converting layouts of non-planarized MMICs is presented. Furthermore, the EM simulation setup and required preprocessing for passive and active circuits is treated.

Chapter 3 first describes the developed EM simulation RF probe tip models of the CM probe and the GGB probe. Next, the corresponding calibration standards on the manufacturers ISS are presented along with the verification of the RF probe tip models by comparing measurement and simulation results of the calibration standards.

Chapter 4 first investigates the performance of the existing RF contact pads as included in the process design kit (PDK) of the Fraunhofer IAF. Subsequently, the occurred parasitic effects are investigated in detail in terms of EM simulations and analytical investigations. Based on the acquired insight, optimized pad structures for packaging and measurement purposes are developed, realized and verified by measurement.

Chapter 5 discusses the principle of on-wafer calibration which is intended to remove the influence of the RF probe to pad interconnect from the measurement results. Evaluation of the existing calibration structures shows that the calibration accuracy is distorted by several parasitic effects such as mode excitation inside the substrate and coupling to adjacent structures. Following the EM based investigations on these parasitic effects, an optimized, electrically robust calibration substrate was developed and characterized.

Chapter 6 analyses the influence of different contact pad structures and calibration methods to the measured circuit performance. Depending on the chosen pad, there is a strong influence on the overall circuit behaviour.

Furthermore, RF probe crosstalk influences the measurement results especially for lossy circuits. To counteract these deviations a technique to correct the crosstalk in measurement results based on EM simulations is developed and successfully applied.

2 MMIC technology and EM simulation setup

The technology, which was used for realizing the test circuits for verification of the investigations in this work, was provided by the Fraunhofer Institute for Applied Solid State Physics (IAF) located in Freiburg, Germany. In this technology metamorphic high electron mobility transistors (mHEMTs) are grown on 4-inch gallium GaAs wafers. For the present work, two technologies with gate lengths of 50 nm [Leu+07] and 35 nm [Leu+13], were used. All transistors are encapsulated in a benzo-cyclo-butene (BCB) layer with a permittivity of $\varepsilon_r = 2.65$, which reduces the parasitic gate capacitance and enhances the RF performance. The 50 nm technology realizes transistors based on a $In_{0.8}Ga_{0.2}As/In_{0.53}Ga_{0.47}As$ composite channel. Due to the very low source contact resistance R_s of 0.15 Ωmm the internal transistor performance is accessible and transit frequency f_t and maximum oscillation frequency f_{max} of 375 GHz are achieved, with a maximum transconductance $g_{m,max}$ of 1800 mS/mm. For the 35 nm process the same $In_{0.8}Ga_{0.2}As / In_{0.53}Ga_{0.47}As$ composite channel is used but scaled down. The achieved transit frequencies f_t and f_{max} are 515 GHz and >1000 GHz, respectively. Compared to the 50 nm technology the maximum transconductance $g_{m,max}$ is increased by almost 40 % to 2500 mS/mm.

The passive technology features comprise several structures such as transmission lines and capacitors, which are based on up to three metal layers. The first metal layer (MET1) with a thickness of 350 nm and a resistivity of 0.1 Ω/□ is used to realize a CPW transmission line environment. Compared to microstrip transmission lines the advantage of CPW transmission lines are low crosstalk and lower radiation. Furthermore, when using active components, CPW has the advantage of direct connection to the ground potential without the use of vias [Sim01]. The second metal layer (MET2) is only available in the 35 nm technology and is separated from MET1 by a 1.4 µm thick BCB layer. The metal thickness and conductivity are identical to MET1. Using MET2 in conjunction with a MET1 ground plane allows to realize thin film microstrip (TFMS) transmission lines. The third metal layer (METG), a galvanic deposited metal with a thickness of 2.7 µm and a resistivity of 8 mΩ/□, is used, among other things, to suppress the excitation of unwanted modes on the CPW transmission lines by realizing airbridges connecting the ground planes above the signal line. The MMICs are passivated using a 250 nm and a 80 nm thick silicon nitride (SiN) layer in the 50 nm and 35 nm technology, respectively. This SiN layer is also used for realizing the metal-insulator-metal (MIM) capacitors by separating METG to MET1 and MET2 achieving a capacitance of 225 $\mathrm{pF/mm^2}$ and 800 $\mathrm{pF/mm^2}$ for the 50 nm and 35 nm process, respectively.

Fig. 2.1 shows the layer stack for different applications of the available layers of the 50 nm technology. For minimizing the loss of transmission lines MET1 can be enhanced with METG to form a thick transmission line. To realize the contact between MET1 and METG the passivation layer SiN has to be opened, see Fig. 2.1a. This technique is used at frequencies below 200 GHz since the geometrical constraints of the CPW transmission lines at higher frequencies would conflict with the minimum gap and feature size of the process. Another application of METG is the realization of airbridges (Fig. 2.1b) to suppress unwanted odd-mode excitation on the CPW transmission line or to route DC connections above signal lines. The realization of MIM capacitors is shown in Fig. 2.1c, showing the separation of MET1 to METG by a layer of SiN. Resistors are realized using a thin film layer of nickel-chromium (NiCr) with a sheet resistance close to 50 Ω/□. The resistive layer is located below the BCB and therefore an opening has to be etched in order to connect MET1 with the NiCr layer, see Fig. 2.1d.

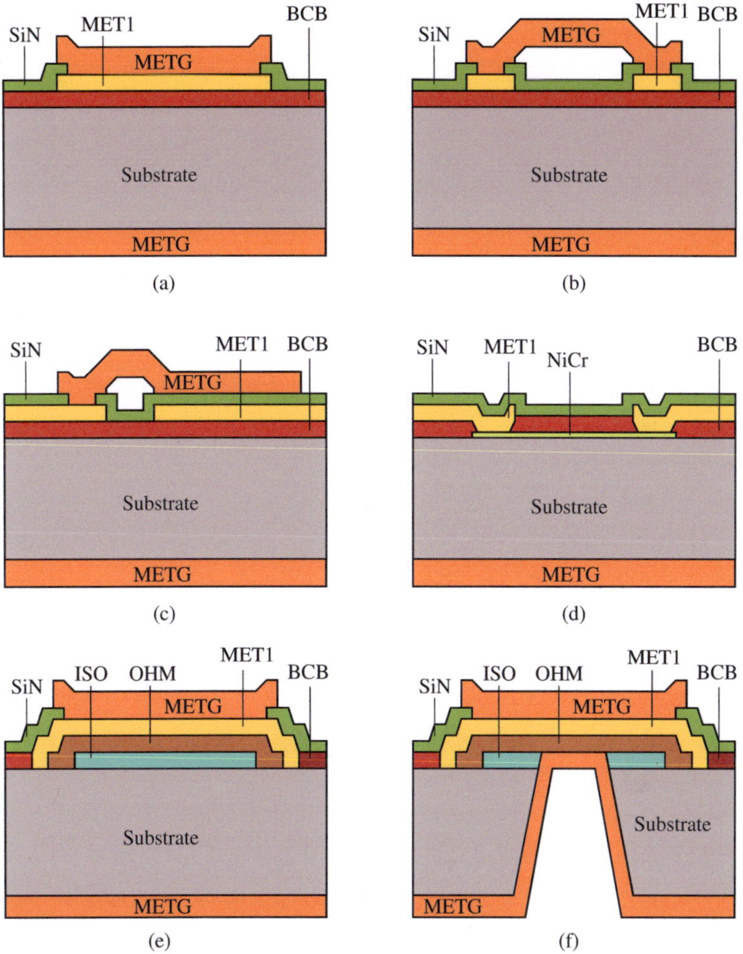

Figure 2.1: Layer stack of the Fraunhofer IAF process for different applications. (a) thick metallization (b) airbridge (c) MIM capacitor (d) NiCr thin film resistor (e) bond pad (f) via hole.

To realize contact and bond pads the thick metal stack from Fig. 2.1a is further enhanced with the conductive OHM and ISO layers, as shown in Fig. 2.1e. The ISO layer is the mesa layer of the MMIC process which is contacted with the thin OHM layer. The compound of these layers results in a good adhesion and therefore high mechanical stability of the pad, which is necessary for wire bonding or flip chip packaging. Vias, which are primarily used for substrate mode suppression, are structurally enhanced on the top-side by the bond pad layer stack. After the frontside processing is finished the wafers are thinned to 50 µm and the vias are etched from the backside, before backside metallization is applied. The resulting layer stack is shown in Fig. 2.1f.

A scanning electron microscope (SEM) picture shows the passive features of the technology in Fig. 2.3a. The picture shows a detailed view of a common MMIC design consisting of bond pad, series and parallel capacitors as well as airbridges and vias. Finally, after all process steps are completed the wafers are diced by laser cutting and the individual dies are stored and shipped in gel-paks®.

2.1 Electromagnetic field simulation of MMICs

To be able to simulate the true EM behaviour of whole MMICs an accurate 3D EM model of the corresponding circuit is required. Compared to printed circuit boards (PCBs), which have a simple layer stack consisting basically only of the substrate and the metal, MMICs have a much more complex layer stack as already discussed in section 2. While most silicon technologies make use of chemical mechanical polishing (CMP) to have each layer at a defined height, so called unplanarized MMICs are common in III-V technologies such as GaAs or indium phosphide (InP). In unplanarized circuits no intermediate CMP is performed to level each layer. Therefore, structures of the same layer are at different heights, depending on which layers are used underneath. This is exemplarily shown in Fig. 2.2 for the METG layer of the Fraunhofer IAF process, which results in several different heights depending on the application e.g. capacitor or airbridge. Therefore, such circuits cannot simply be imported into commercially available EM simulation tools, which expect a layer at the same height across the whole MMIC. The challenges and solutions of importing unplanarized MMIC layouts are discussed in the following section.

2.1.1 CAD model generation of passive circuits

To allow the import of unplanarized MMICs into the EM simulation software, a sophisticated conversion tool was developed in this work.
For passive circuits, i.e. circuits without active components such as transistors or varactors, the layout requires no preprocessing before conversion to the EM simulation suitable format. The tool uses the circuit layout of the MMIC and slices a single layer into multiple layers depending on the underlying layers. For this task the layer of interest is compared by geometrical boolean operations with the underlying layers. The generated layers are each on a defined height and can therefore be imported into the EM simulation tool using standard importing tools. For simplification of this conversion, the layer BCB is modelled directly on the substrate regardless of any potential underlying layers. Therefore, the thin layers ISO and OHM, which are actually below the BCB layer are modelled as recess layers into the substrate. This reduces the number of different layer combinations involving METG from 18 to 5. Fig. 2.2 shows the remaining combinations and corresponding height of the METG layer above the substrate.
Similar considerations were made for the layers SiN and MET2 (in case of the 35 nm technology) whose absolute heights are also dependent on the underlying layers. The final converted EM circuit models consist of up to 18 and 38 layers for the 50 nm and 35 nm technology, respectively. A qualitative comparison between converted EM simulation model and a SEM picture of the corresponding structure is shown in Fig. 2.3. The corresponding

cross-section of the converted layout is depicted in Fig. 2.3c and shows the resulting layer stack of a serial MIM capacitor in detail.

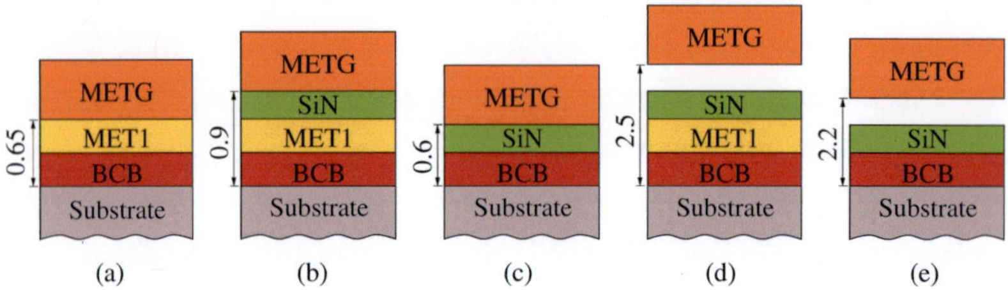

Figure 2.2: Possible layer combinations involving METG (a) thick metal (b) MIM capacitor (c) METG routing on SiN (d) airbridge routing above MET1 (e) airbridge routing without MET1 (dimensions in μm).

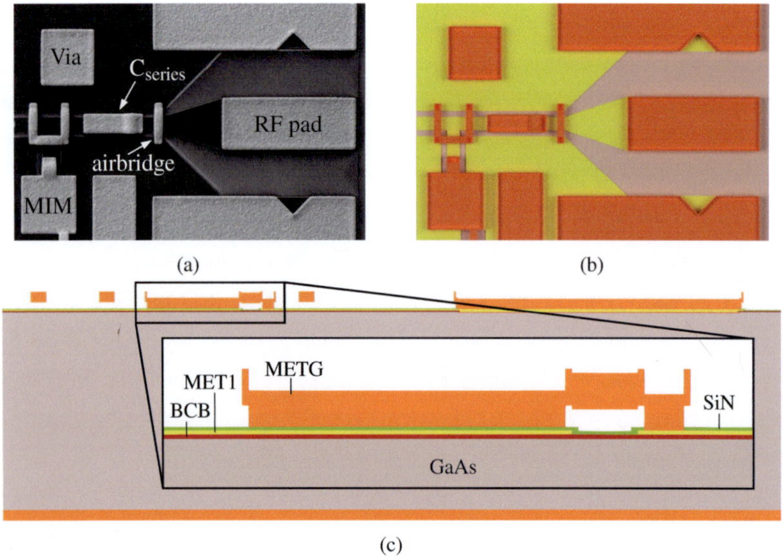

Figure 2.3: Comparison of (a) SEM picture [Tes+08], (b) converted EM simulation model (the layers SiN and BCB are hidden) and (c) cross-section of the same structure.

2.1.2 Simulation setup for passive circuits

Even though commercially available EM tools offer several ways of signal excitation, initial investigations indicated that the integrated methods may lead to severe deviations between measurement and simulation. The origin of this lies in the dimensions of the RF contact pad and the RF probe, which are in the WR3 frequency range both in the dimension of the wavelength, leading to complex interaction. To comprehend this behaviour it is essential to include a model of the RF probe into the EM simulation. Therefore, the imported circuits are excited with partial models of the RF probe, contacting the circuit identical to the real life measurement setup. The signal excitation is performed by WG ports at the coaxial conductor of the RF probe model, where a pure mode can be assumed. While the upper frequency is limited in a measurement setup by the VNA extension modules and the used waveguides, the simulations are not bound by these limitations. However, the upper simulation frequency is limited by the excitation of higher order modes in the coaxial cable of the RF probe model. Analytically, with an inner conductor diameter of 80 µm and an outer conductor diameter of 270 µm, the next higher mode starts to propagate at 380 GHz but the effects start to arise at lower frequencies. Therefore, to keep a certain frequency reserve, the maximum simulation frequency was set to 360 GHz. The lower frequency limit was chosen at 170 GHz, which is a compromise between an increased frequency range and clarity of the resulting figures. Furthermore, extending the frequency range may lead to inaccurate results since the actual RF probe geometry of the corresponding bands may differ. To represent the measurement setup environment in the simulation, electric boundaries were chosen at the z_{min} plane, which is a good approximation of the metal chuck of the probe station. An open add space boundary condition was chosen at the remaining planes, representing the surrounding free space. The simulation was done in CST Microwave Studio® (MWS) using the frequency domain solver with a tetrahedral mesh. For the investigated structures with the very thin layers of the MMIC models and the rather large geometries of the RF probe models, the tetrahedral mesh is superior to the hexahedral mesh, due to its geometrical flexibility. The mesh size limits were set to a maximum step of 35 µm for the RF probe model and 12 µm for structures of the MMIC model, which is 1/6 and 1/20 of the shortest wavelength, respectively. Simulations showed that a lower step size does not improve the accuracy of the results but significantly increases the required simulation time.

2.1.3 Preparation of active MMICs for EM simulation

The work-flow for passive as well as active DUTs is depicted in Fig. 2.4. While passive circuit layouts require no specific preparation, active circuit components such as transistors or varactors underlie quantum mechanic effects and can therefore not be simulated by commercially available EM software tools, which are based on solving the Maxwell equations. This requires to subdivide the circuit of interest into two parts. The active components are removed from the layout before converting and importing them into the EM simulation tool using the conversion process described in section 2.1. Once the layout is imported WG ports are added at the reference planes of the active to passive interfaces. These ports are able to excite a pure CPW mode at the interface plane, which is a good approximation of the field distribution at this interface and was already successfully used in [Die+12] for simulating the parasitic elements of a transistor shell.

The active components of the MMIC require DC connections to provide the supply voltages and currents, which is usually done by DC contact pads in the upper and lower area of the chip. To contact these pads, DC probes are used. Because these DC probes are close to the RF probe tips during a measurement, it is necessary to also include them into the simulation in order to reproduce the measurement setup as close as possible and therefore get precise simulation results. The frequency range, mesh and boundary conditions were chosen identical to the passive simulation setup.

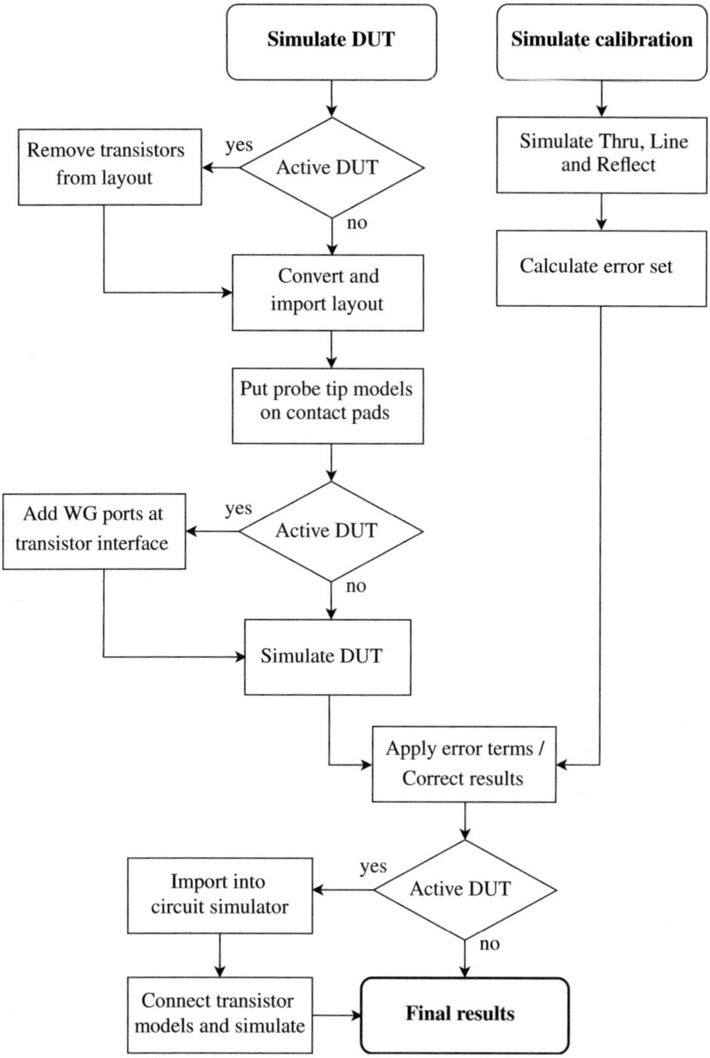

Figure 2.4: Flowchart of the procedure for simulating and calibrating active and passive MMICs including probe tip models.

After the EM simulation is completed, the results are imported into the circuit simulator software Keysight Advanced Design System (ADS). Models of the active components are then attached to the ports and by importing the previously calculated error terms the RF probe models are de-embedded and the actual circuit behaviour is simulated.

2.2 Conclusions of chapter 2

In this chapter first the Fraunhofer IAF indium gallium arsenide (InGaAs) technology is presented, which is used for realization of the various MMICs required for the verification. Due to transit frequencies of at least 375 GHz, this technology builds a solid foundation for mmW and sub-mmW circuit design. However, the non-planarized layer stack does not allow the use of standard import tools for EM simulation software. The resulting layer heights depending on the underlying layers are discussed in detail. This issue was resolved by using boolean operations and slicing one layer to multiple layers each with a defined height. The development of a specific software tool allowed to convert existing layouts into a suitable format for importing tools of common EM simulation software. Therefore, MMIC designs can be imported with only little preprocessing, including all structural details and a full 3D EM verification can be carried out.

3 Electromagnetic field simulation of RF probes

Key to simulating the RF probe influence onto the resulting behaviour of the DUT is a precise EM model of the used RF probe. Since the type of RF probe has an influence on the DUT, two RF probes of different manufacturers and distinct practical realizations were investigated. Each of these probes has a corresponding ISS, which was also modelled and simulated. In analogy to the measurement, the simulated calibration standards were used to calculate a corresponding error set, which is used to correct the simulations and remove the influence of the probe model [Mül+17d].

3.1 Modelling of the RF probes

A photograph of both used probes has already been shown in Fig. 1.4. Common for both probes is the WR3 waveguide flange, where the measurement equipment is connected. The waveguide is bent at an angle of approximately 45° and mounted to the probe body. A broadband transition from waveguide to coaxial cable is embedded inside the probe body. At the probe tip the coax mode is converted into a CPW mode for the three contact pins.
Compared to the actual probe tip, which consumes less than 1 % of the overall spatial probe volume and the size of the MMICs, the overall probe dimensions, including the probe body and the waveguide connector, are tremendous in terms of wavelengths at mmW frequencies.

However, simulations indicated that the probe body and waveguide connector have no significant influence as long as the radiation behaviour of the DUT is of no interest, i.e. no antennas are embedded on the MMIC.
To reduce the size of the EM simulation volume, and therefore being able to solve the problems with the available computational power, only a small part of the probe was modelled. This part comprises the actual probe tip as well as approximately 1 mm of the remaining probe.
Since neither Cascade Microtech Inc. nor GGB Industries Inc. provide computer-aided design (CAD) models of their probes, the models presented below are based on a few available publications, microscopic photographs and educated assumptions on the internal structures.

3.1.1 Cascade Microtech Infinity WR3 probe

The model for the CM probe was generated using the description given in [Cam+07; Cas04; Cas] and microscopic photographs. The cross-section of the final model is shown in Fig. 3.1a.
In reference to the contact plane the coaxial cable is bent to approximately 45°, ground down at the end and the dielectric is removed to a certain depth. To contact the DUT the CM probe uses a polyimide (PI) membrane with two metal layers embedding a coax to microstrip and microstrip to CPW transition. The distance between the three wafer contacts defines the pitch of the RF probe, whereby the RF probes used in this work have a pitch of 100 μm. This membrane is attached to the ground down coaxial cable by means of soldering. After attaching the membrane, the probe tip is mechanically stabilized with epoxy glue. The excess epoxy is afterwards ground off, to allow sight on the probe contact pins with a microscope from above. This is essential for proper probe placement during measurements. In order to suppress unwanted propagation along the outside of the coaxial cable, the conductor is surrounded by an absorptive material. A comparison of the final model to microscopic photographs at various angles can be seen in

Fig. 3.2. The lowest operating frequency of the real probe is limited by the cutoff frequency of the basic mode and the upper frequency limit is set by the cutoff frequency of the next higher mode. For a WR3 waveguide these frequencies are at 173.5 GHz and 347.1 GHz. The model of the probe tip, however, does not depend on the WR3 characteristics since the waveguide is not part of the model. Therefore, the model can theoretically be used for simulations down to DC. The upper frequency limit of the probe tip model is limited by the excitation of higher modes in the coaxial cable, which start to occur at 390 GHz. However, it has to be kept in mind, that using the probe tip model outside of its specified frequency range might lead to results which differ from the actual results when a suitable RF probe is used. This is due to a change of probe tip geometry depending on the frequency band and therefore other parasitic effects, which might occur.

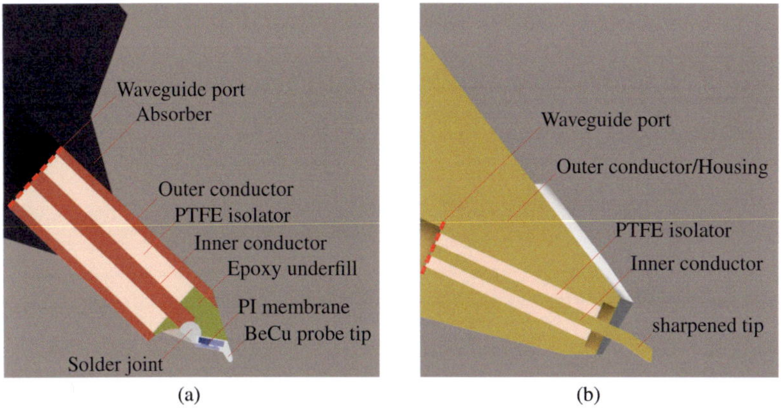

Figure 3.1: Cross-section of the CAD RF probe models (a) CM Infinity Probe (b) GGB Picoprobe.

(a)

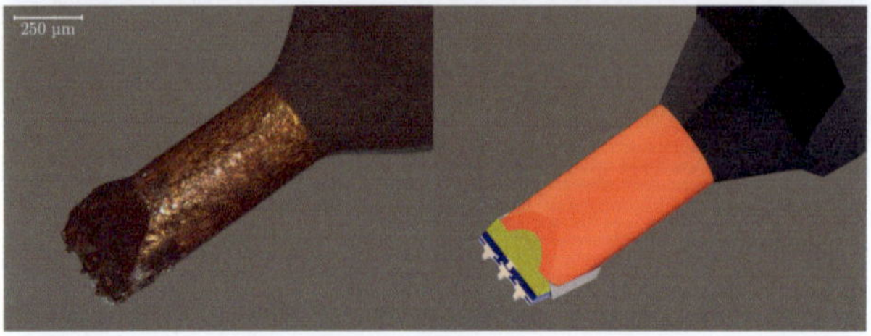

(b)

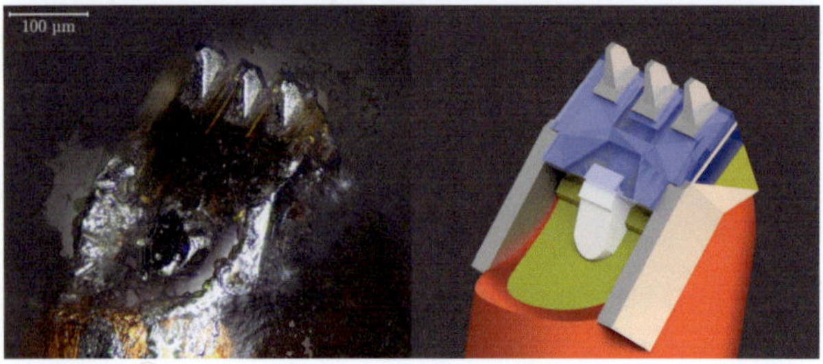

(c)

Figure 3.2: Microscopic photograph of a CM probe and the CAD model (a) sideview (b) diagonal from above (c) diagonal from below.

3.1.2 GGB Picoprobe WR3 probe

For the GGB Picoprobe only very limited information is given by the manufacturer [GGB] and therefore the model is mostly based on several microscopic photographs and assumptions about the internal construction. The cross-section of the final model is shown in Fig. 3.1b. After the waveguide to coax interconnect the GGB probe uses a square coaxial cross section, which is milled into the split-block body of the probe. The inner conductor is isolated with polytetrafluoroethylene (PTFE) shell as dielectric against the split-block case, which acts as ground reference. The two mirrored parts of the probe body are assembled by closely arranged screws to achieve a minimum gap between them. A gold-plating is applied to the probe body to minimize the electrical losses of the mmW signal. Towards the actual probe tip the split-block is narrowed down and a slot is cut to facilitate mounting of the ground contacts. Depending on the pitch, different spring loaded Beryllium-Copper wing-like triangular plates with a sharp contact area are soldered directly into the cut of the split-block. The excess solder is also included in the simulation model. The probes used in this work have a pitch of 80 µm. The center contact pin consists of the extension of the inner conductor and is sharpened to ensure a small scrub area with reliable contact to the DUT. The drawback of the probe design is the short distance from the actual probe tip to the probe body, which limits the use of DC probes for bias supply, due to geometrical constraints. This restricts the use of the probe for verification of either passive MMICs without DC supply or large MMICs with a minimum lateral distance of 250 µm between RF pad and DC pad. The final model compared to microscopic photographs of three different perspectives is shown in Fig. 3.3.

(a)

(b)

(c)

Figure 3.3: Microscopic photograph of a GGB probe and the CAD model (a) sideview (b) diagonal from above (c) diagonal from below.

3.2 Verification of the proposed models

To apply the same calibration procedure used in measurements also for the RF probe tip simulation, it was necessary to create CAD models of the employed ISS calibration standards. The manufacturer's recommendation for the WR3 frequency range is to use the CM ISS 138-357 for the CM probe and the GGB Industries Inc. CS-15 ISS for the GGB probe. Overview pictures of both ISS are shown in Fig. 3.4. The simple layer stack of both ISS consisting only of the Alumina substrate and a 3 µm thick gold metal layer simplifies the CAD modelling drastically. Both substrates comprise the standards open, short, match and thru as well as several lines of different lengths. This allows the use of various calibration techniques such as SOLT, LRRM, TRL or mTRL. For this work only the TRL calibration was used, due to its simplicity (see section 1.2.2).

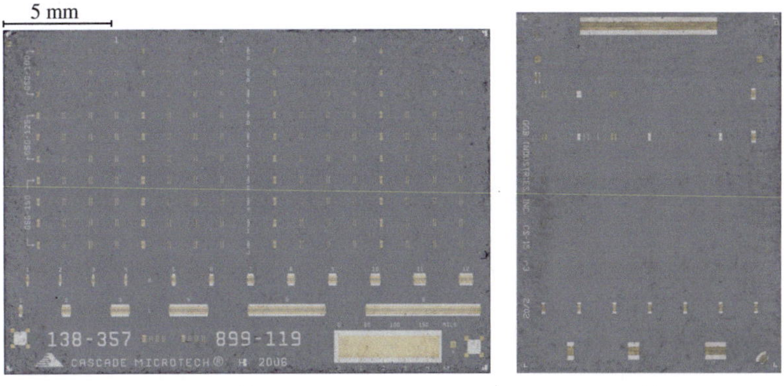

(a) (b)

Figure 3.4: Photograph of the (a) CM ISS 138-357 and (b) GGB CS-15 ISS.

3.2.1 Cascade Microtech ISS calibration substrate

An overview of the CM ISS 138-357 calibration substrate is shown in Fig. 3.4a and close-up microscopic photographs of the individual standards are shown in Fig. 3.5.
The ISS provides groups of standards for RF probe pitches of 100 μm, 125 μm and 150 μm. For this work the standards suitable for 100 μm pitch RF probes were used and schematic pictures showing the dimensions of the standards are shown in Fig. 3.6. The standards open, short and load are based on the same basic contact pad layout. While the open standard has no connection to the ground areas, the short standard is connected by a 25 μm wide connection in the gold layer. The load standard is realized by two parallel 100 Ω resistors consisting of a 18 μm wide resistive layer with a sheet resistance of 45 Ω/□, see Figs. 3.6a-c. The thru standard, shown in Fig. 3.6d, is designed to have a characteristic impedance close to 50 Ω and a length of 135 μm. The layout of the line standards differs from the thru standard by larger ground planes and additional absorptive areas at the outer edge of the ground planes. These absorptive areas are intended to suppress field propagation along the edge of the standard. For TRL calibration of the WR3 frequency range the recommended line standard, shown in Fig. 3.6e, has a length of 250 μm. This yields to an optimal TRL frequency range from 62 to 497 GHz.

3.2.2 GGB Industries Inc. calibration substrate

The overview of the GGB CS-15 calibration substrate is shown in Fig. 3.4b and microscopic photographs of the utilized standards are shown in Fig. 3.7. While the GGB CS-15 ISS basically comprises the same standards as the CM ISS 138-357, the layout is carried out in a different manner. The layout of the CS-15 standards does not distinguish the RF probe pitch and uses the same standards for a pitch range from 40 to 150 μm. Schematic drawings of the standards with detailed dimensions are shown in Fig. 3.8.

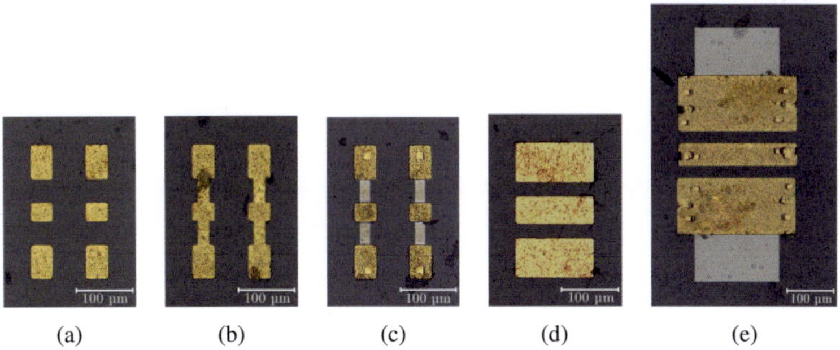

Figure 3.5: CM ISS 138-357 standards (a) open, (b) short, (c) load, (d) thru and (e) line.

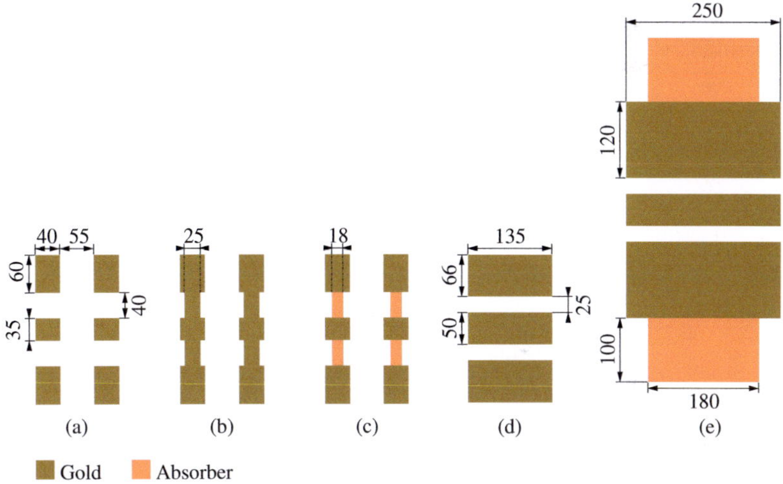

Figure 3.6: Schematic representation (dimensions in µm) of the CM ISS 138-357 standards (a) open (b) short (c) load (d) thru (e) line.

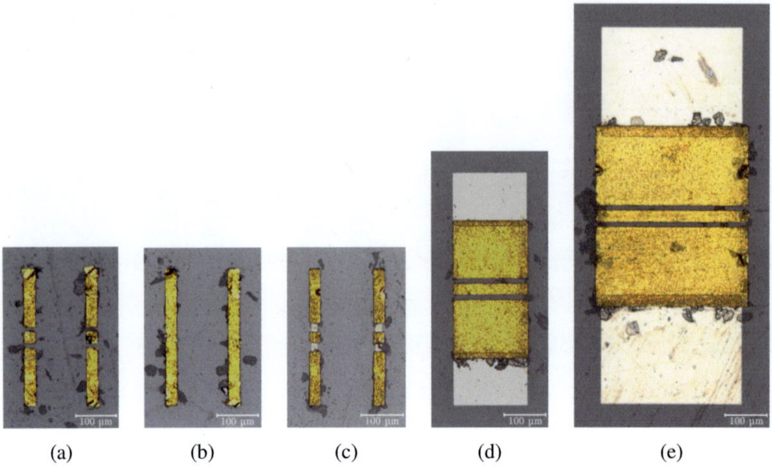

Figure 3.7: GGB ISS CS-15 standards (a) open, (b) short, (c) load, (d) thru and (e) line.

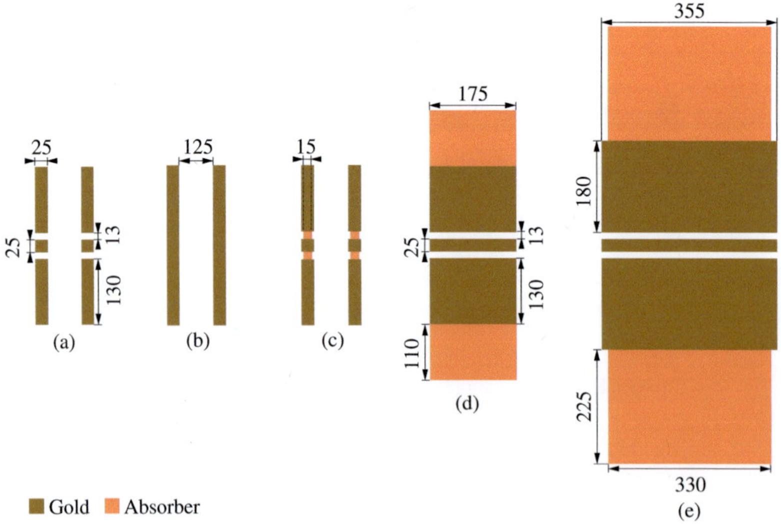

Figure 3.8: Schematic representation (dimensions in µm) of the GGB CS-15 ISS standards (a) open, (b) short, (c) load, (d) thru and (e) line.

As for the CM ISS, the GGB CS-15 standards for open, short and load are built on the same basic contact pad layout with the short and load standard derived from the open standard, see Figs. 3.8a-c. The distance between two opposite standards is 125 µm, which is more than two times the distance of the CM calibration standards and is intended to minimize RF probe crosstalk. The area of the inner conductor is 25 µm × 25 µm and therefore only 45 % of the CM ISS contact area. The dimensions of the inner conductor and the gap size to the ground planes is optimized to a characteristic impedance of 50 Ω. Therefore, the same dimensions of the inner conductor width and gap width to the outer conductor are also used for the thru and line standards, as shown in Figs. 3.8d and e. The thru and line standards also have a resistive sheet with a resistivity of 60 Ω/□ at the edge of the ground planes, which is intended to suppress field propagation along the outer edges. The length of the available thru and line standard for the WR3 frequency range are 175 µm and 355 µm, respectively.
The length difference of both lines fulfils the TRL criteria in the frequency range from 39.7 to 317.3 GHz, therefore the calibration quality at the upper WR3 band edge is degrading.

3.2.3 Comparison of measured and simulated calibration standards

To verify the deduced RF probe models, measurement and simulation results of the calibration standards were compared. The simulation results of the thru, line and open standard were used to calculate a TRL error set, which is then used to make the error correction for the simulations. Verifying the calibration on the same standards, which were used for calibration is only of limited significance, because the used standards are partially mapped to the expected behaviour. However, there are several standards available on the ISS which were not used during the calibration: The short, match and open in air standard are not required for the TRL calibration and are therefore available for the verification.

Both RF probes, CM probe and GGB probe were measured and simulated using their corresponding calibration substrate with reference planes shifted directly to the RF probe tips. For direct comparison, the results of measurement and simulation of the ISS 138-357 standards contacted with the CM probe are shown in Figs. 3.9 and 3.10. The reflection coefficient S_{11} shown in Fig. 3.9 for the frequency range from 200 to 330 GHz shows a good agreement between measurement and simulation. Generally, all Smith diagram charts shown in this work are restricted to the frequency range 200 to 330 GHz and do not use the extended frequency range from 170 to 360 GHz. This is necessary to keep the curves of simulation results comparable to the measurements. Comparing the results of the short and open standard in Fig. 3.9 shows that the simulated phase of S_{11} has less variation, which may be caused by a slightly different RF probe position between measurement and simulation. However, it can be seen that measured and simulated standards all include an inductive component and are therefore located in the upper half of the Smith chart. The investigations carried out in appendix A indicate that this behaviour depends on the used calibration substrate and not on the used RF probe.

Comparison of the transmission coefficient S_{21} in Fig. 3.10, also shows a good agreement between measurement and simulation, even at low magnitude levels. For the sake of clarity the line standards are shown separately in Figs. 3.10a and 3.10b. As can be seen, simulation and measurement are almost identical and the results follow the expected behaviour of a transmission line with increasing loss and decreasing phase towards higher frequencies. The S_{21} results of the reflection calibration standards are shown in Figs. 3.10c and 3.10d. Simulation and measurement are generally in very good agreement down to magnitudes of −40 dB, which is close to the noise level of the measurement setup. It is notable that even at these low magnitudes the phase of S_{21} is predicted accurately by the EM simulations.

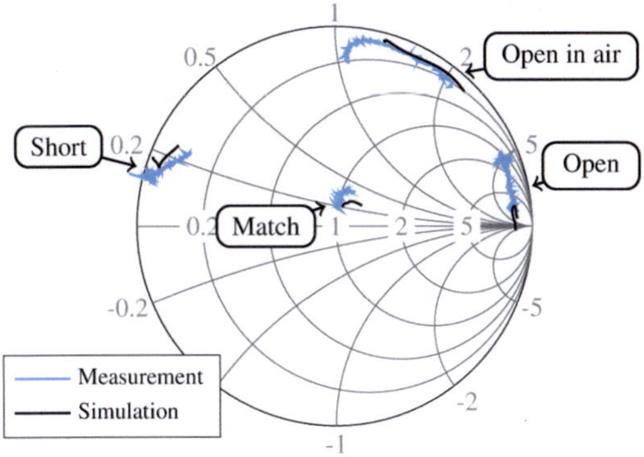

Figure 3.9: Measured and simulated S_{11} of CM ISS 138-357 standards contacted with a CM probe in the frequency range 200 to 330 GHz [Mül+17d].

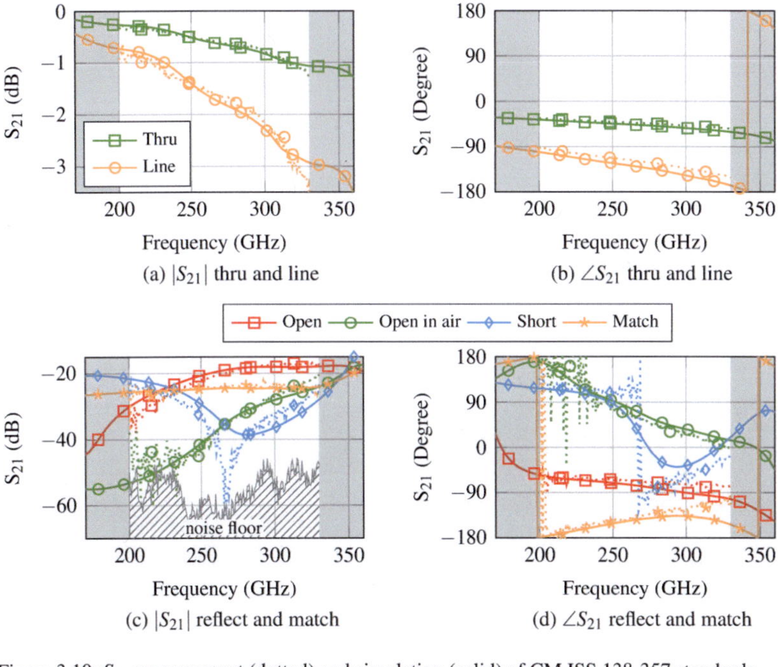

Figure 3.10: S_{21} measurement (dotted) and simulation (solid) of CM ISS 138-357 standards contacted with CM probes [Mül+17d].

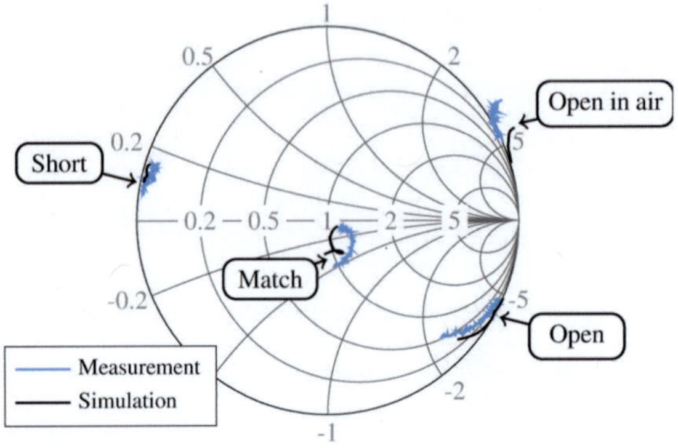

Figure 3.11: Measured and simulated S_{11} of GGB CS-15 ISS standards contacted with a GGB probe in the frequency range 200 to 330 GHz.

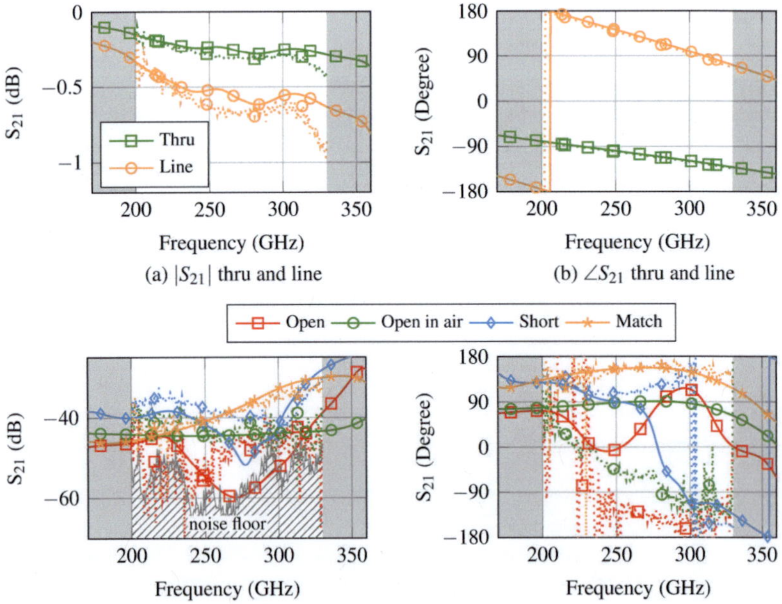

(a) $|S_{21}|$ thru and line

(b) $\angle S_{21}$ thru and line

(c) $|S_{21}|$ reflect and match

(d) $\angle S_{21}$ reflect and match

Figure 3.12: Measured (dotted) and simulated (solid) S_{21} of GGB CS-15 standards contacted with GGB probes.

The verification of the GGB probe with the corresponding CS-15 ISS is shown in Figs. 3.11 and 3.12 for the input reflection coefficient S_{11} and the transmission coefficient S_{21}, respectively. Analogue to the verification of the CM probe model, a very good agreement between measurement and simulation is observed. Compared to the CM probe and ISS results, Fig. 3.11 shows that the calibrated match, open and open in air standard of the GGB setup have less inductance, which shifts the match and open standard into the capacitive region of the Smith chart.
The transmission coefficient S_{21} is evaluated in Fig. 3.12. Compared to the CM ISS the calibrated results of the line standards result in less loss and lower phase variation, see Figs. 3.12a and 3.12b. Generally, this behaviour is predicted very precisely by the RF probe tip simulation. The evaluation of the reflect and match standards is shown in Figs. 3.12c and 3.12d for the magnitude and phase of S_{21}, respectively. Again a good prediction even of the phase is achieved for magnitudes above -40 dB whereas lower magnitudes vanish into noise.
Up to this point the evaluation was focused on the RF probe tip model with its corresponding ISS. This, however, is prone to errors due to a co-dependency of errors in the RF probe tip model to errors in the calibration standards, i.e. it is possible that an error in the RF probe tip model is compensated by another error in the model of the calibration standard. To investigate the accuracy of the RF probe tip models, further simulations and measurements were performed. Therefore, a setup with RF probe and ISS of different manufacturers was investigated, i.e. the CM probe was calibrated using the GGB CS-15 ISS and the GGB probe was calibrated on the CM ISS138-357. Furthermore, this evaluates whether the reflect, transmission and crosstalk behaviour depends on the RF probe or the utilized ISS. The measured and simulated results are presented in appendix A. While some deviations between measurement and simulation occur, the overall agreement is very good, proving the quality of the RF probe tip models. The evaluation of the reflect standards shows that the location in the Smith chart

is to a first order dependent on the used ISS and not on the RF probe. Evaluating the crosstalk shows a lower coupling of the GGB Picoprobe independent of the used ISS.

3.3 Conclusions of chapter 3

In this chapter, the two used RF probes manufactured by Cascade Microtech Inc. and GGB Industries Inc. were discussed in detail along with the generation of the corresponding EM models and the calibration procedure. While both RF probes address the same frequency range, they differ significantly in their physical realizations. This also affects the necessary effort in creating the corresponding EM simulation models. The CM probe using a coaxial conductor with an attached structured membrane embedding the RF probe tips has several degrees of freedom, which required an extensive amount of microscopic pictures and design iterations to achieve satisfying results. The GGB probe on the other hand uses a split block realization with a simpler geometrical structure. The ground pins are directly soldered to the body of the RF probe. A sharpened extension of the inner conductor of the coaxial line, which is milled into the split block body, is used as the center pin of the RF probe. This simple physical realization also relaxed the EM modelling and satisfying results were achieved after few design iterations. The verification of both RF probes using the manufacturers calibration substrates showed a very good agreement between measurement and simulation. Furthermore, verification using the competitors calibration substrate verified the robustness of the probe models as well as the models of the calibration standards. Compared to the CM probe the GGB Picoprobe showed less crosstalk with up to 25 dB difference for the open in air measurement at the same RF probe distance. Generally, the EM simulation results are in very good agreement to the measurements and precisely predict the different effects even at low magnitude levels for crosstalk investigations, proving that the occurring parasitic effects are precisely replicated by the EM simulations. Generally, the very good agreement between measurement and simulations allows to investigate the influence of the RF probes on actual MMICs, which is carried out in the following chapters.

4 Evaluation and optimization of RF pad structures

The task of the MMIC contact pads is to connect the internal circuitry to the outer world. Ideally a contact pad shows transparent behaviour in terms of RF performance, i.e. the internal circuit characteristics are not altered by the pad and the true performance is accessible from the outside. Measurements in the WR3 frequency range however show that depending on the type of contact pad a large influence on the overall circuit performance is present. This raises the suspicion that there are parasitic effects in the available contact pads, which were up to now unobserved. This chapter first analyses the behaviour of the existing RF pads in the Fraunhofer IAF PDK, as available in March 2016, in terms of RF performance and their qualification for packaging purposes. Using the EM probe tip models presented in section 3.1 allowed to precisely comprehend the occurring deviations by evaluating the corresponding E-fields. Following this, improved pad structures are designed, realized and verified by measurements. By utilizing the RF probe tip models it was furthermore possible to optimize the contact pads within only one design iteration and achieve satisfying results with resonance-free behaviour, low loss and good matching up to 330 GHz. Compared to the classical contact pad optimization, where the actual behaviour is only visible after realization and measurements of suitable test structures, this technique saved a significant amount of time and money.

4.1 Evaluation of the existing RF pads

4.1.1 Description of the existing RF pads

To evaluate the existing PDK RF pads, simple test-structures were designed, embedding two pads at the input and output, which are connected by a 170 µm long CPW transmission line. The three different RF pads of the PDK, suitable for WR3 frequencies, were evaluated and are shown in Fig. 4.1. Each of these pads is designed for a specific application.
The pad shown in Fig. 4.1a is intended for both measurement as well as packaging purposes. Therefore, the dimensions are chosen to fit the requirements imposed by the available packaging technology, i.e. wire bonding or flip chip. The actual contact area of the pad makes use of the full layer stack shown in Fig. 2.1e with both available metal layers. With a size of $30 \times 75\,\mu m^2$ the dimensions of the inner conductor are large enough to allow wire bonding or flip chip packaging. Therefore, this pad configuration is referred to as 'bond-pad' in the following discussion. The gap size of the inner conductor to the ground planes sets the characteristic impedance of the CPW mode, which is usually intended to be close to 50 Ω. Given the substrate height of 50 µm, the gap size to achieve a characteristic impedance of 50 Ω is approximately 60 µm, leading to a parasitic microstrip mode, which is a result of the improper gap to substrate height ratio and the metallized backside of the MMICs.
To achieve a dominant coplanar mode, the gap was chosen to 27 µm, resulting in a characteristic impedance of approximately 40 Ω. Simulations showed that, if exited with a CPW mode, this has only minor influence on the overall performance. After the contact area, a taper using the first metal layer is embedded, converting the pad dimensions to the intended CPW environment with a ground-to-ground spacing of 14 µm and an inner conductor width of 7.4 µm, resulting in a characteristic impedance of 50 Ω.

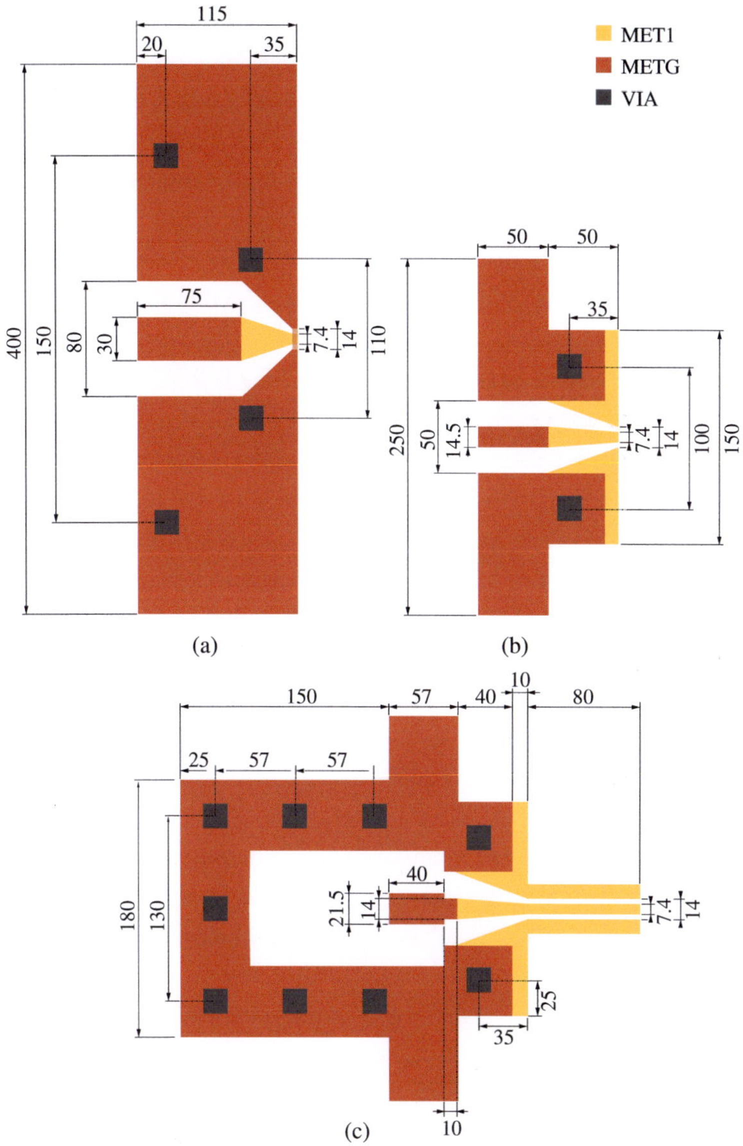

Figure 4.1: Schematic layouts (dimensions in μm) of the (a) bond-pad (b) measurement-pad and (c) characterization-pad.

Due to unsymmetrical probe placement the excited field may not be a perfect CPW mode and therefore an airbridge connecting both ground planes is added at the narrow end of the taper to suppress potential odd-mode excitation. A second pad for WR3 frequencies, which is only intended for measurement purposes, is shown in Fig. 4.1b. In the design of the pad, the dimensions were not bound by the packaging requirements and it was possible to optimize the characteristic impedance to approximately 50 Ω. The dimensions depicted in Fig. 4.1b show that compared to the bond-pad, the ground-to-ground spacing as well as the width of the inner conductor are narrower. The basic idea behind this downscaling is to prevent the excitation of a parasitic microstrip mode. Again, a taper is added after the contact area to adapt to the CPW environment. The narrow dimensions make it difficult to package the circuit with standard packaging technology and hence this pad is referred to as 'measurement-pad', since it is predominantly used for circuit verification by on-wafer measurements.
The third pad, shown in Fig. 4.1c, is predominantly used for high accuracy characterizations of transistor test structures [See+15] and is therefore referred to as 'characterization-pad'. The dimensions were not restrained by packaging requirements, resulting in a characteristic impedance close to 50 Ω. In addition to the taper, a 80 μm long CPW transmission line is added to ensure a pure CPW mode at the reference plane for on-wafer calibrations. In contrast to the bond-pad and measurement-pad, this pad is backed by a closed via ring connected to ground, which acts as an electric shield to the remaining circuit. Furthermore, the shielded environment in the area close to the probe contact position minimizes the influence of the surrounding structures. The drawback of this pad is the large required chip-area and the fact that, due to the via ring, it is not feasible to connect it with short bond wires. This is however crucial for broadband and low loss interconnects at mmW frequencies.
Microscopic photographs of the corresponding test-structures used to verify the characteristics of each pad are shown in Figs. 4.2a - c.

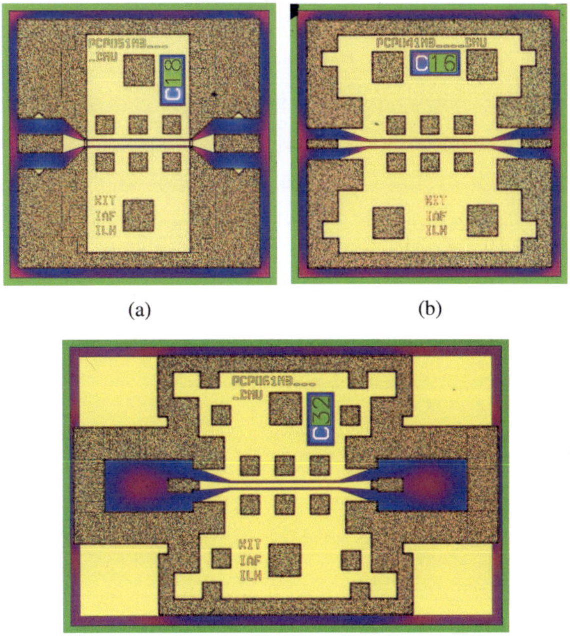

(a) (b)

(c)

Figure 4.2: Photograph of the contact pad test structures for (a) bond-pad, (b) measurement-pad and (c) characterization-pad.

4.1.2 Evaluation of the existing RF pads

All three test structure MMICs were measured and EM simulated including RF probe tip models in conjunction with an ISS TRL calibration, which shifts the reference plane to the probe tips. A 3D rendering of the simulation setup including the probe tips is exemplarily shown for the bond-pad test structure in Fig. 4.3. A fine tetrahedral mesh was chosen for the simulation which resulted in approximately 420,000 mesh cells per structure, including the probe tip models. Both RF probes presented in section 3 were used for the investigations, i.e. CM probes and GGB probes were used in the measurement as well as in the simulation setup. The calibrated results of the test structures are shown in Figs. 4.4 - 4.6. In addition to the measurements and EM probe tip simulations, the test structures were also EM simulated with WG port excitation at the same position as the probe contact, which is the common excitation in EM simulations of CPW structures. With a single CPW mode excited at the contact position, this estimates the maximum achievable performance of the corresponding structures. Evaluation of the measurement and simulation results is carried out in the following sections. Due to the symmetry and the passive nature of the DUTs, only the input reflection coefficient (S_{11}) and forward transmission coefficient (S_{21}) are considered in the following discussions.

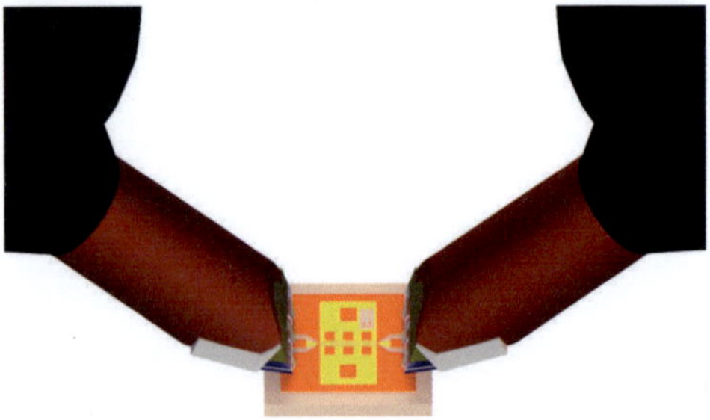

Figure 4.3: Exemplary CAD rendering of a bond pad test structure contacted with CM probe tip models.

PDK bond-pad evaluation

The resulting reflection coefficients S_{11} of the bond-pad test structure are shown in Fig. 4.4a. While there are small differences between measurement and probe tip simulation, the overall agreement is very good over the observed frequency range. At frequencies below 270 GHz the measured reflection coefficient, when contacted with the CM probe is even better than predicted by simulation. The reflection coefficient measured and simulated with the GGB probe is also in very good agreement over the full frequency range. However, compared to the WG simulation, the measured and probe tip simulated reflection coefficient S_{11} is severely increased.

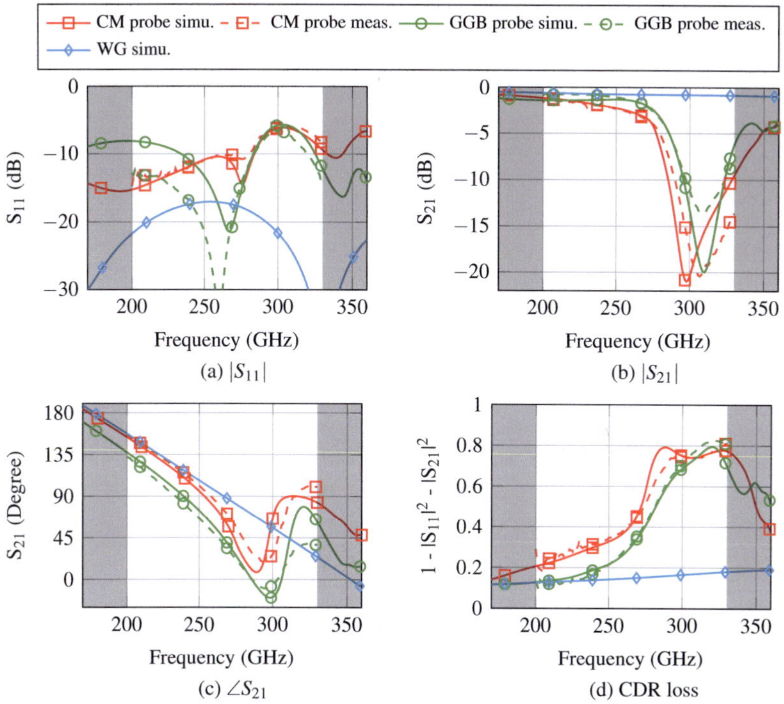

Figure 4.4: Measurement and simulation of the PDK bond-pad test structure.

Evaluation of magnitude and phase of the transmission coefficient S_{21} is performed in Figs. 4.4b and 4.4c, respectively. At frequencies around 300 GHz a large deviation from the ideal WG simulation is visible for both probes. A resonance is clearly visible in measurement and simulation for both probe tips. The transmitted signal is strongly attenuated, resulting in amplitudes of only -14 dB and -20 dB for the GGB probe and CM probe, respectively. The phase of S_{21} also strongly deviates form the expected behaviour predicted by the ideal WG simulation. Fig. 4.4d shows the conductor, dielectric and radiation (CDR) loss of the structure as calculated by [Poz12]:

$$\text{CDR Loss} = 1 - |S_{11}|^2 - |S_{21}|^2 \tag{4.1}$$

At frequencies around 200 GHz the measured and simulated CDR loss for both probe tips is close to the prediction of the ideal WG simulation. However, towards higher frequencies the loss of the WG simulation only slowly increases whereas the measurements and probe tip simulations show a strong increase with a maximum loss at approximately 315 GHz. At this frequency 80 % of the incident energy is either radiated or absorbed in metal and dielectric losses, which altogether results in a drastic decrease of the transmission coefficient S_{21}. Above the specified frequency range the overall CDR loss is predicted to decrease again.

PDK measurement-pad evaluation

Evaluating the measurement and simulation results of the measurement-pad, the test structure shows comparable behaviour to the bond-pad test structure. The reflection coefficient S_{11} shown in Fig. 4.5a shows that the measured and probe tip simulated behaviour is again worse than predicted by the ideal WG simulation. While the matching is below -10 dB for almost all frequencies with both probes, the matching of the GGB probe is generally slightly better over the full band. The transmission coefficient S_{21}, shown in Figs. 4.5b and 4.5c, shows a resonance at approximately 295 GHz, which is comparable to the resonance of the bond-pad test structure but not as severe in the magnitude influence. A minimum S_{21} magnitude of approximately -6 dB and -4.3 dB for the CM probe and GGB probe are observed, respectively. The increased transmission loss consequently also increases the overall loss of the structure, as shown in Fig. 4.5d. At 200 GHz the GGB probe shows almost identical loss to the ideal WG simulation whereas the loss of CM probe is slightly higher. Both probes stay below 70 % loss, which is about 13 % lower compared to the bond-pad test structure.

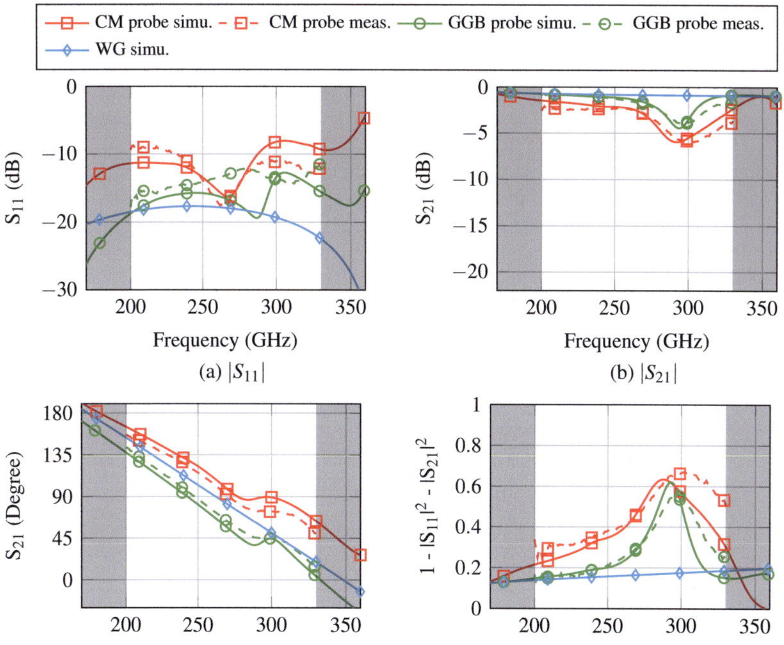

(a) $|S_{11}|$ (b) $|S_{21}|$ (c) $\angle S_{21}$ (d) CDR loss

Figure 4.5: Measurement and simulation of the PDK measurement-pad test structure.

PDK characterization-pad evaluation

The characterization-pad test structure shows the lowest deviations from the WG simulation results and moreover generally of all PDK pad test structures. Even though the input reflection coefficient S_{11} shown in Fig. 4.6a is above the predicted ideal behaviour, the resulting curves show no resonant behaviour. Evaluating the transmission coefficient S_{21} in Figs. 4.6b and 4.6c confirms the resonance free operation in the WR3 frequency range. The S_{21} magnitude shows less than 2.2 dB insertion loss up to 330 GHz with a linear decreasing phase. The overall loss of the structure is shown in Fig. 4.6d and is below 40 % over the full frequency range. Especially the measurement and simulation employing the GGB probe are in good agreement to the loss predicted by the WG simulation.

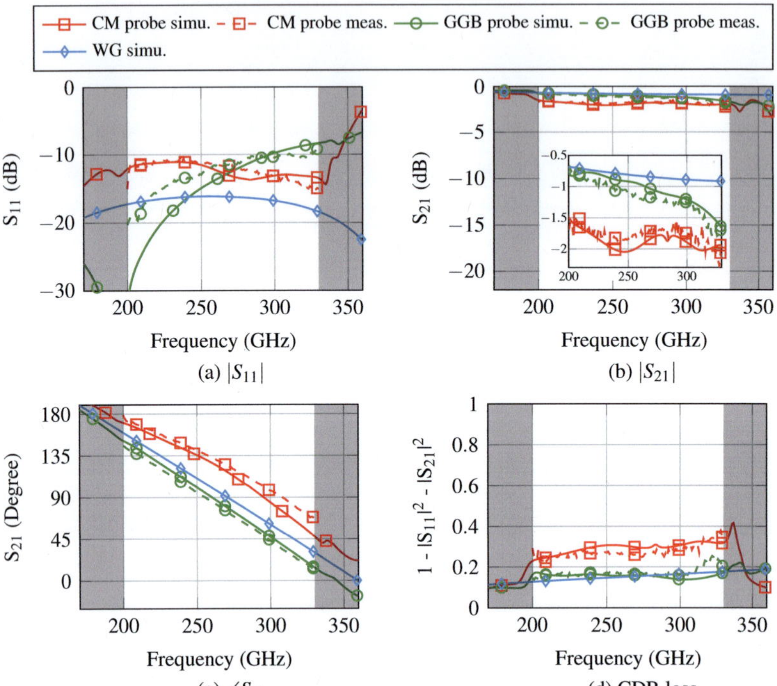

(a) $|S_{11}|$

(b) $|S_{21}|$

(c) $\angle S_{21}$

(d) CDR loss

Figure 4.6: Measurement and simulation of the PDK characterization-pad test structure.

In conclusion, only the characterization-pad offers a good performance over the full WR3 frequency range with a maximum loss of 2.2 dB. The measurement-pad and bond-pad show a severe resonance around 300 GHz, with a maximum loss of 6 dB and 20 dB, respectively. This limits the use of these pads to frequencies below 250 GHz. The observed resonance is not visible if the circuits are simulated using the WG port excitation, but are precisely predicted if the RF probe tip models are used for excitation.

4.2 Analysis of parasitic modes in RF-pads

As discussed in the previous section, two of the PDK RF pads show severe deviations from the expected behaviour, which may lead to malfunction of the MMIC. The origin of these deviations is subject to the investigations of the following section. The good agreement between measured and simulated S-parameters allows the conclusion that the occurring parasitic effects are precisely reproduced by the EM simulations including the RF probe tip models. This allows furthermore to evaluate the E-field of the structures at several frequencies and investigate the origin of the observed deviations. After this, an analytic model is developed, which is able to predict the resonance frequency dependent on the dielectric properties of the substrate and the via placement in the vicinity of the RF pad structure.

4.2.1 Evaluation of EM fields at the resonance frequency

Special focus in the evaluation of the E-fields was on the resonance occurring around 300 GHz, which diminishes the transmission coefficient S_{21}. The E-field components were evaluated exemplarily for the bond-pad test structure, which shows the strongest deviations. So called E-field monitors, which calculate the complex E-field tensor at the frequency of interest, were simulated and evaluated. Since measurements and simulations employing the CM probe showed stronger distortions, the following investigations are based on the use of this RF probe. Fig. 4.7 shows a z-plane cross-section of the test structure along with an overlay of the E-field magnitude. Fig. 4.8 shows the same simulation but the cross-section is chosen in the y-plane. As can be seen, the electric-field (E-field) at 200 GHz is well concentrated inside the MMIC substrate below the contact pad.

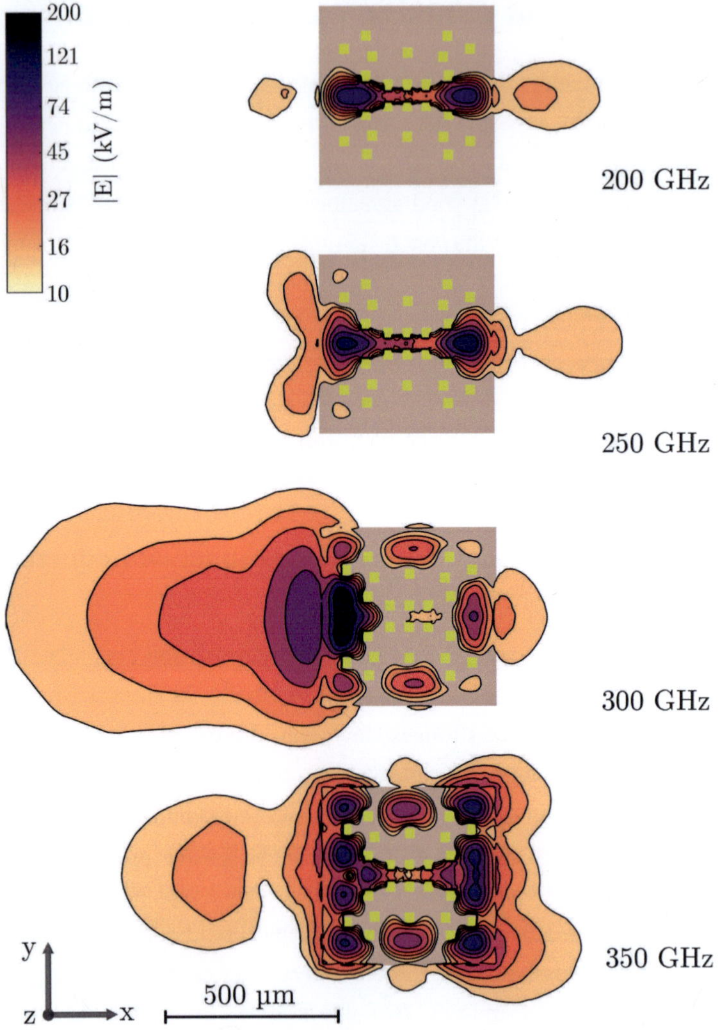

Figure 4.7: Top view of a z-plane cross-section at the center of the MMIC showing the E-field magnitude for different frequencies of the PDK bond-pad test structure (feed left).

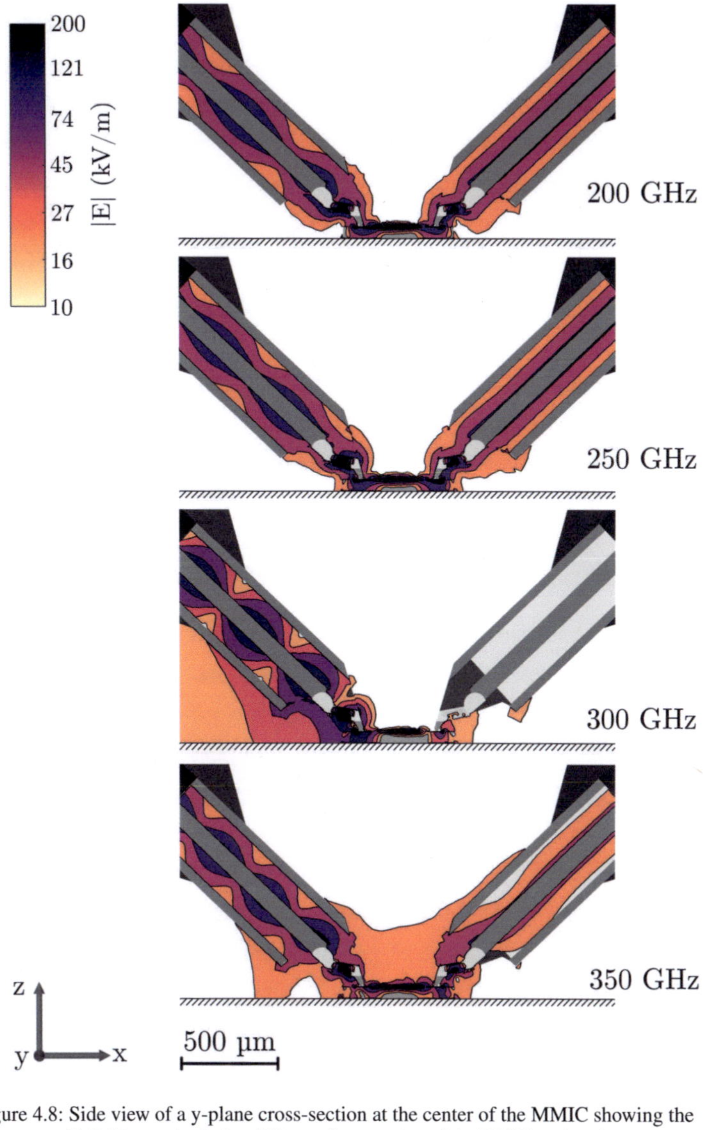

Figure 4.8: Side view of a y-plane cross-section at the center of the MMIC showing the E-field magnitude for different frequencies of the PDK bond-pad test structure (feed left).

Starting at 250 GHz the E-field starts to radiate outside of the MMIC. At 300 GHz the E-field strongly radiates towards the input RF probe and a microstrip mode is present between the outer conductor of the RF probe and the metal chuck, which is represented by a ground plane. The decrease in transmission coefficient S_{21} around 300 GHz is clearly visible by a vanishing magnitude of the E-field inside the coaxial conductor of the output RF probe, which is on the right hand side of Fig. 4.8. At frequencies around 350 GHz the microstrip field components under the RF probe decrease and the field components inside the output RF probe increase again. However, multiple parasitic modes are excited inside the MMIC in the volume between the vias, as can be seen in Fig. 4.7. These results clearly show, that the number and placement of vias have a large influence on the overall circuit behaviour at mmW frequencies.

4.2.2 Analytical investigation of parasitic modes

While the EM simulations using the RF probe tip models provide an accurate prediction of the pad resonance frequency, the model generation is complex and the simulations themselves are time consuming. An analytically based prediction on the other hand is able to achieve instant results and allows a simple understanding of the occurring effects. A simplified drawing of a conventional contact pad is shown in Fig. 4.9. The pad consists of a CPW topology with the inner conductor and the ground planes being the landing area of the RF probe or the bond wires. After the contact area, the inner conductor and ground planes are tapered down to the intended dimensions of the CPW transmission line used for connecting the internal circuitry. Since the backsides of the MMICs are metallized it is necessary to introduce vias for substrate mode suppression. The evaluation of the E-fields in section 4.2.1 indicates that top metal, bottom metal and the vias form a resonant cavity, which is shown as a dashed overlay in Fig. 4.9. The dimensions sketched in Fig. 4.9 can be considered only as a first order approximation of the resonator's dimension, since the vias do not form a closed conducting wall and electric field components enter the area between the vias, extending the effective size of the resonator. Furthermore, the pad is usually not directly placed at the edge of the chip but has a certain offset, which is required to facilitate dicing of the MMICs. This offset leads to fringing fields, comparable to a parallel plate capacitor, effectively increasing the resonator volume and decreasing the resonance frequency.

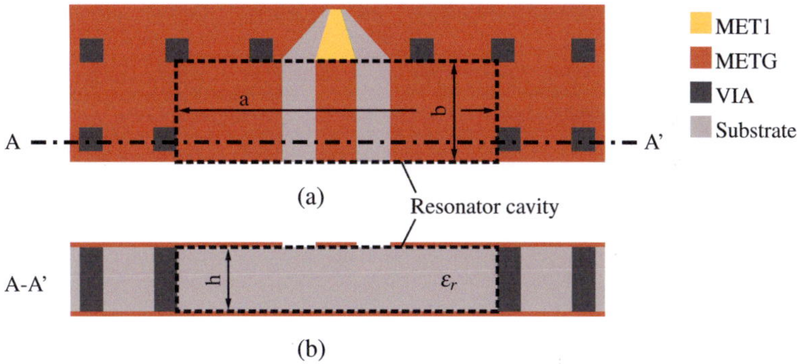

Figure 4.9: Schematic representation of a conventional contact pad (a) top-view (b) cross-section. The dashed overlay shows the dimensions of the resonator cavity.

To calculate the resonance frequency of the cavity, a simplified model was developed and is shown in Fig. 4.10. A cuboid with the dimensions of the resonator cavity shown in Fig. 4.9 and the dielectric constant ε_r of the substrate, was used for the analytical investigation. The boundary condition was chosen to electric (no transverse electric fields) at all faces that are in contact with metal i.e. top and bottom of the MMIC and at faces that are in direct contact with vias. At the interface between substrate and air, the boundary of the cuboid is chosen to magnetic (no transverse magnetic fields), which is a good approximation for substrates with a high dielectric constant [ZL08], as common in semiconductor technologies.

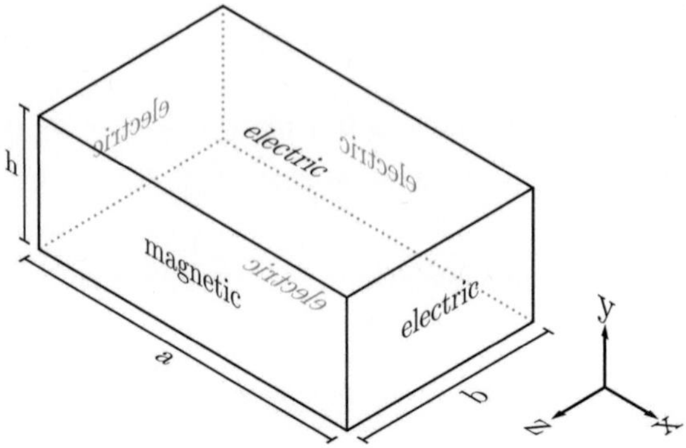

Figure 4.10: Model of the resonator cavity showing the orientation and boundary conditions as well as the dimensions.

The dominant mode (lowest resonance frequency) of such a cuboid can be derived from the transverse E-field E_t of a rectangular waveguide, which is given by [Poz12]:

$$\bar{E}_t(x,y,z) = \bar{e}(x,y)[A^+ e^{-j\beta_{mn}z} + A^- e^{j\beta_{mn}z}] \tag{4.2}$$

with $\bar{e}(x,y)$ being the transverse variation of the mode, A^+ and A^- the amplitudes of the forward and backward travelling waves and β_{mn} the propagation constant of the corresponding mode with order m and n. At $z = 0$ the electric field is shorted by the closely arranged vias and therefore the boundary condition is set to:

$$\bar{E}_t(x,y,0) = 0 \tag{4.3}$$

resulting in the equivalence $A^+ = -A^-$, therefore simplifying (4.2) to:

$$\bar{E}_t(x,y,z) = -2jA^+ \bar{e}(x,y)\sin(\beta_{mn}z) \tag{4.4}$$

Due to the magnetic boundary at $z = b$ the transverse magnetic field has to fulfil the following condition:

$$\bar{H}_x = \frac{\bar{e}(x,y)A^+}{Z_{\mathrm{TE}}} \cos(\beta_{mn} b) = 0 \tag{4.5}$$

with $Z_{\mathrm{TE}} = E_x/H_y$. This results in the nontrivial solution $\beta_{mn} b = \pi(n+1/2)$. Therefore, the resonance frequency of the m, n-th mode can be calculated with [Poz12]:

$$f_{mnl} = \frac{c}{2\sqrt{\mu_r \varepsilon_r}} \sqrt{\left(\frac{m}{a}\right)^2 + \left(\frac{n + 1/2}{b}\right)^2 + \left(\frac{l}{h}\right)^2} \tag{4.6}$$

where the dominant mode TE_{100} has a resonance frequency calculated by:

$$f_{100} = \frac{c}{2\sqrt{\mu_r \varepsilon_r}} \sqrt{\left(\frac{1}{a}\right)^2 + \left(\frac{1/2}{b}\right)^2} \tag{4.7}$$

Evaluating (4.7) shows that for a typical pad configuration where the dimensions fulfill the condition $a > b > h$, the dominant mode is dependent on the width a and length b of the cavity and independent of the substrate height h. The obtained solution gives a fast estimation, whether or not a pad configuration is prone to parasitic modes in the frequency range of interest and can be adapted to various substrates without effort. For GaAs substrates with a dielectric constant $\varepsilon_r = 12.9$ the resulting critical frequencies for a variety of the parameters a and b are shown in Fig. 4.11. Considering that using a conventional RF probe with a pitch of 100 µm will generally require the width of the pad to be at least $a = 200$ µm, it can be seen that especially when going into the high mmW frequency range around 300 GHz, a proper pad design has to be carried out. When designing a contact pad, it has to be kept in mind that a certain frequency reserve is necessary, since the deterioration caused by the resonance has a specific bandwidth, depending on the quality factor of the resonance. Initial comparison to existing pad configurations showed that the resonance frequencies can be predicted using (4.7) with errors below 20 %. Yet, the simple analytical model gives no insight on the quality factor of the resonance and therefore on the frequency reserve that has to be considered in the pad design. As already discussed, to precisely predict the resonance and overall pad behaviour, the use of EM field simulation in conjunction with the corresponding RF probe models is inevitable.

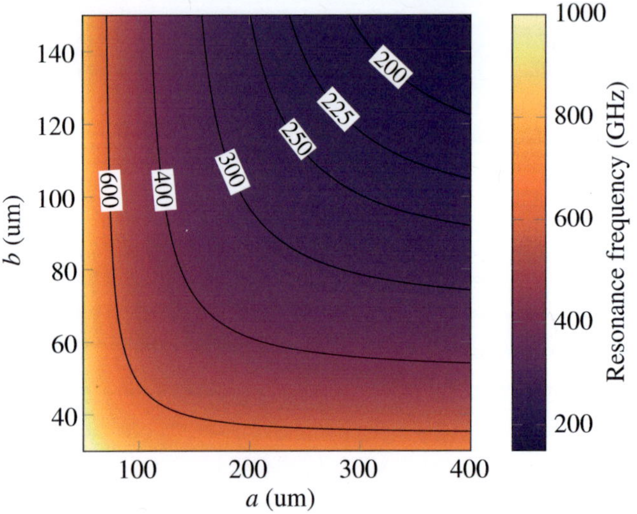

Figure 4.11: Resonance frequency of the dominant mode depending on the parameters a and b of the RF contact pad calculated with (4.7) [Mül+18c].

4.3 Optimized bond pad design

After the comprehensive analysis of the resonant pad behaviour this section discusses the optimization of the RF pad structures. The main design goal is to achieve resonance free operation with good matching and low loss in the frequency range of interest, i.e. in this work up to 330 GHz. The dimensions given by the simplified evaluation of the previous section are considered as a starting point and corresponding RF pad designs were carried out according to the design rules of the Fraunhofer IAF 50 nm and 35 nm technology. These designs were first validated by EM simulations including the RF probe tip models and the most promising candidates were realized and characterized by measurements.

4.3.1 Design rule conformal pad structures

Up to this point the investigations on the pad performance depending on the size of the resonator cavity were based on a simplified analytical model and assumptions that were not necessarily in agreement with the design rules. To fulfil the design rules, the pad requires a certain minimum distance between the metal layers and the edge of the MMIC. This leads to an extension of the substrate at the outside of the pad, resulting in fringing fields and an additional enlargement of the effective resonator volume. In addition the process comprises several dielectric layers which are not included in the analytical model such as BCB or SiN. To predict the real behaviour as close as possible and to simulate effects that may be suppressed by the simplified model, four test structures were designed which fulfil the design rules of the MMIC process. Parameters that had to be considered were for example distance between adjacent vias, minimum feature and minimum gap size as well as thickness of the dielectric and metal layers. These test structures were translated as described in section 2.1 and imported into CST MWS, including all layers and vias. To achieve accurate predictions, the CM probe models presented in section 3.1 were used in the simulations, together with the corresponding error set for calibration of the results.

4.3.2 Variation of via placement in pad design

To investigate the influence of the quantity and position of vias in the contact pad design, four different configurations shown in Fig. 4.12 were evaluated [Mül+18c]. When arranging the positions of the vias it has to be kept in mind that probing or wire bonding directly on a via may lead to a crack in the surface and therefore destruction of the pad. The pad configuration shown in Fig. 4.12a has no additional vias in the contact plane and can therefore be probed with arbitrary pitch and packaged without taking care of underlying vias. Additional vias were added in Figs. 4.12b and 4.12c which limits these pads to RF probes with a pitch ranging from 50 to 150 μm and 50 to 100 μm, respectively. The pad shown in Fig. 4.12c is identical to the bond-pad of the PDK. The configuration shown in Fig. 4.12d is optimized for RF probes with a pitch of 100 μm, which allows to add additional vias in-between the contact area of the RF probe tips and represents the closest possible via spacing within the used pad layout.

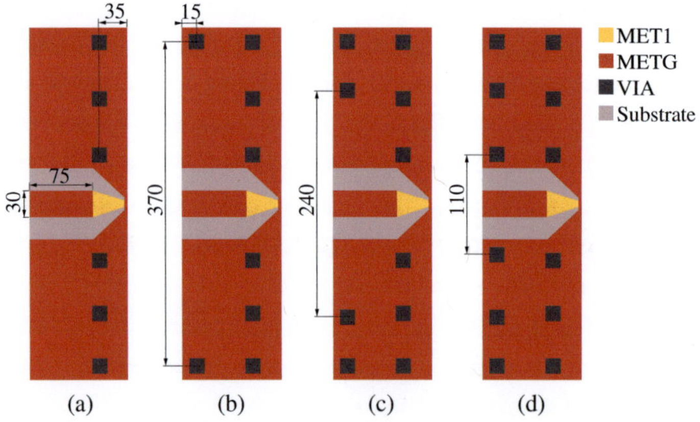

Figure 4.12: Different via placement for a RF probe pitch of (a) A: 50 to 200 μm (b) B: 50 to 150 μm (c) C: 50 to 100 μm (d) D: 100 μm. (Dimensions in μm).

Inserting the approximate resonator dimensions in (4.7) gives the resonance frequencies of these structures at 337 GHz, 363 GHz and 531 GHz for pad

B, C and D, respectively. Since pad configuration A has no vias which confine the electric field, the boundary assumptions of the analytical equation are not fulfilled and the equation cannot be used. To verify the resonance and overall behaviour of the different pad configurations, test structures analogue to the PDK pad test structures were designed. As an example, a schematic drawing of the test structure for verification of pad configuration D is shown in Fig. 4.13. In addition to the MMIC, the position of the RF probes is sketched showing the contact position of the probe pins. Furthermore, the plane of the cross-section B-B' which is used for evaluation of the E-field distribution in section 4.3.2 is marked. The size of the test structure is $0.5 \times 0.5\,\text{mm}^2$ with an offset of 50 µm between top metal and the chip edge, which ensures that no debris drops on the pad structure during the dicing process. Located on the backside of the chip is a closed metal plane with a lateral distance of 35 µm to the chip edge. In addition to the vias that are embedded in the pad, the MMIC contains additional vias to suppress any potential mode excitation inside the substrate. Especially in the direct vicinity of the CPW transmission line, vias are added to avoid radiation into the substrate.

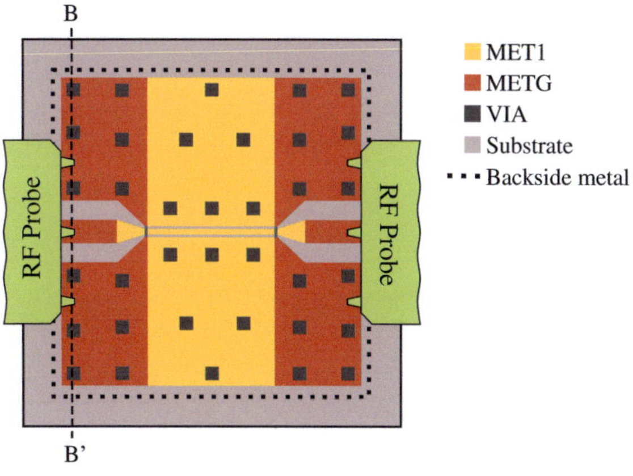

Figure 4.13: Top-view of the test structure for the different pad configurations including part of the RF probe tip.

S-parameter simulation results of different via configurations

Simulations of the different test structures were performed in the frequency range from 110 GHz to 360 GHz and calibrated with an ISS TRL calibration error set of the corresponding frequency range. The results of the calibrated S-parameters are shown in Figs. 4.14a-d. To evaluate the distortion introduced by the different pad configurations, the internal CPW line connecting both pads was also simulated stand-alone without any connected pads.

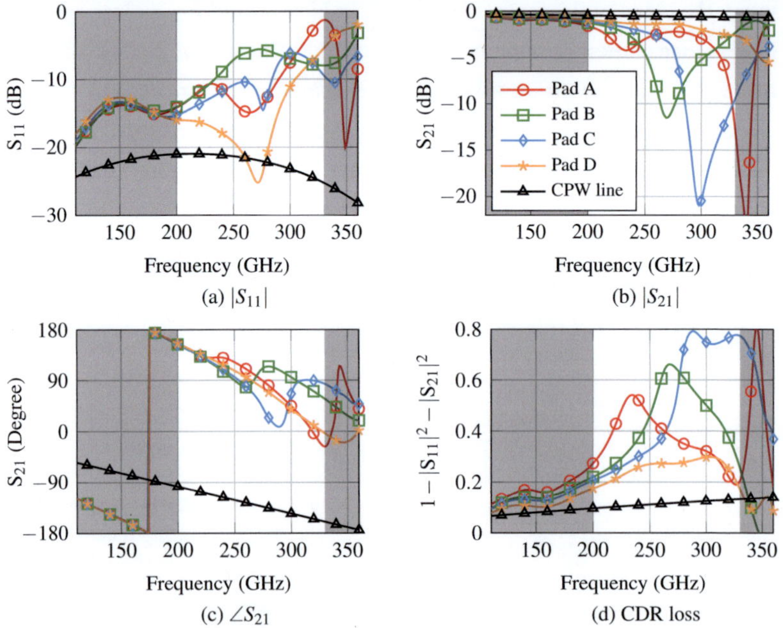

Figure 4.14: S-parameter simulation results of different pad via configuration test structures and the internal line connecting the pads [Mül+18c].

Evaluation of the simulated results in Fig. 4.14a shows, that up to 200 GHz the different pad configurations have almost identical behaviour with a reflection coefficient S_{11} better than -12 dB. Above 200 GHz the behaviour of the different test structures deviates strongly. While pad configuration A,

B and C show a rapid decrease in the reflection coefficient up to −5 dB and worse, pad configuration D is matched below −10 dB up to 305 GHz. Evaluation of the transmission coefficient S_{21} of the different test structures shows a maximum loss of 1.6 dB up to 200 GHz, which occurs for the test structure of pad configuration A, as shown in Fig. 4.14b. However, it has to be kept in mind that this is outside of the specified frequency range for the used RF probe model and actual values might differ. Above 200 GHz the results start to deviate and the impact of the parasitic modes is clearly visible. Pad configuration A, which has no additional vias, has the first resonance at 230 GHz and a second resonance is occurring outside of the specified frequency range towards 350 GHz, leading to local minima in the transmission coefficient S_{21}, as shown in Fig. 4.14b. Similar behaviour can be observed for pad configuration B and C. The pads are based on the same topology but pad C has two additional vias, which reduces the resonator cavity volume and therefore increases the resonance frequency, as can be seen in Fig. 4.14b. At the resonance frequencies 270 GHz and 298 GHz transmission coefficients S_{21} of −11.5 dB and −21 dB are observed for pad configuration B and C, respectively. Pad configuration D, embedding the maximum number of vias, shows no resonance in the specified frequency range with a maximum loss of 3.2 dB. The phase of S_{21} for all structures is depicted in Fig. 4.14c and shows that the structures have generally a linear decrease of phase up to the resonance frequency. At the resonance frequency, the phase of S_{21} is delayed and deviates towards higher frequencies. The resonance free pad configuration D shows a linear decrease in the phase of S_{21} with only minor distortions towards the upper specified frequency limit. Above the specified frequency range the simulation results indicate first signs of deviations from the ideal behaviour. Calculating the overall CDR loss according to (4.1) of the stand-alone CPW transmission line shows the expected behaviour with a monotonic increasing loss towards higher frequencies. Evaluating the calculated loss for pad configuration A shows that both resonances result in a decrease in transmission coefficient

S_{21} but the effect of the second resonance is much stronger, as shown in Fig. 4.14d. The origin of this behaviour are modes of different order, which result in different radiation behaviour, depending on their field distribution. Compared to configuration B the loss of pad C is about 20 % higher and influences a wider bandwidth. Pad configuration D shows the lowest loss of the investigated pads with a maximum loss of 30 % at 305 GHz, as can be seen in Fig. 4.14d.

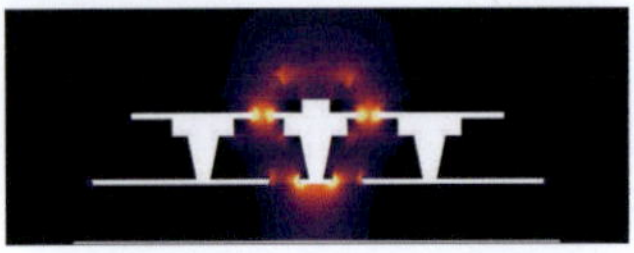

(a) Pad A: E-field magnitude for 200 GHz

(b) Pad A: E-field magnitude for 230 GHz

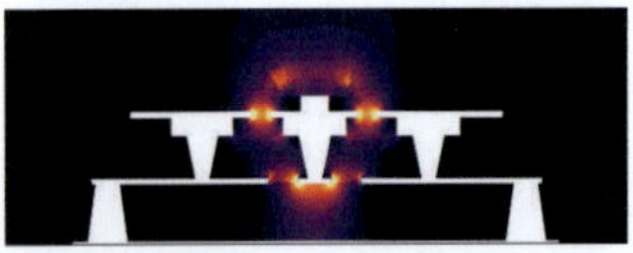

(c) Pad B: E-field magnitude for 200 GHz

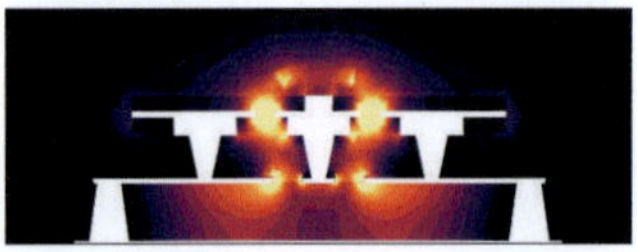

(d) Pad B: E-field magnitude for 270 GHz

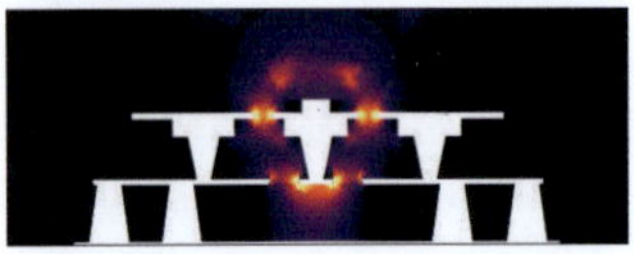

(e) Pad C: E-field magnitude for 200 GHz

(f) Pad C: E-field magnitude for 300 GHz

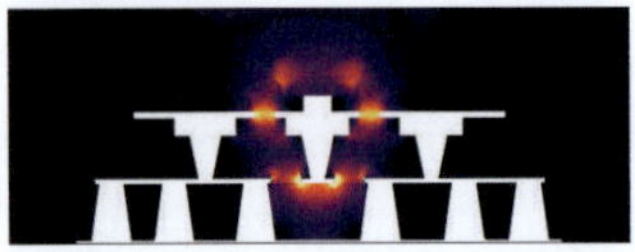

(g) Pad D: E-field magnitude for 200 GHz

(h) Pad D: E-field magnitude for 330 GHz

Figure 4.15: Magnitude of the E-field for all four pad configurations at the cross-section B-B' for 200 GHz and at the corresponding resonance frequency [Mül+18c].

E-field evaluation of different via configurations

The electric field magnitude in the cross-section B-B' (indicated in Fig. 4.13) for the different pad configurations is shown in Fig. 4.15. The white areas indicate zero electric field magnitude, which is the case inside conducting objects, e.g. RF probe tips, vias and metal layers. Generally, there is a moderate to high field concentration in the upper half of the plot, which shows the gap of the internal CPW structure of the RF probe tips. At low frequencies the field distributions of the different pad configurations are almost identical, as can be seen exemplarily at 200 GHz in Figs. 4.15a, c, e and g. The major part of the field is concentrated in the gap between the inner conductor and the ground planes on top of the MMICs, with only low field magnitudes towards the bottom ground plane.

Field distributions of the pad configurations at the corresponding resonance frequencies are shown in Figs. 4.15b, d, f and h. Without any additional vias, pad A shows a strong field concentration at the edges of the top ground planes (Fig. 4.15b) with only low field magnitudes towards the bottom ground plane. With an increasing number of vias, pad B and C confine the electric field and successfully suppress radiation at the edges of the top metal. With increasing frequency, the field starts to concentrate towards the bottom ground plane and microstrip-like behaviour is observed (Figs. 4.15d and f). Especially pad C shows a microstrip-like mode below the inner conductor of the contact pad, which leads to a strong radiation, as discussed earlier. Analysis of the E-field distribution of pad B and C showed, that the field distribution of pad C is comparable to the field distribution of a dielectric resonator antenna, therefore leading to stronger radiation at the corresponding resonance frequency compared to pad B. This microstrip mode is mostly suppressed in pad configuration D (Fig. 4.15h) by the close via arrangement. Therefore, no resonance occurred in the simulated frequency range and the electric field distribution is plotted at the maximum specified frequency of 330 GHz.

4.3.3 Verification of the optimized bond-pad structure

Due to the promising results of pad configuration D this pad was selected as a candidate for the optimized bond-pad. To verify the resonance free behaviour up to 330 GHz the corresponding test structure was realized and measured. A microscopic photograph of the test structure is depicted in Fig. 4.16. The outer dimension of the chip is $0.5 \times 0.5\,\text{mm}^2$. The CM probe as well as the GGB probe were utilized for the verification with a corresponding TRL calibration. The calibrated simulation and measurement results are shown in Fig. 4.17.

Figure 4.16: Chip photograph of the test structure for pad configuration D. The chip size is $0.5 \times 0.5\,\text{mm}^2$.

While the detailed discussion of the PDK bond-pad was already carried out in section 4.1, the results are again shown in Fig. 4.17 to allow direct comparison to the optimized bond-pad. Evaluating the reflection coefficient S_{11}, shown in Fig. 4.17a, shows a dependency on the used RF probe. While the CM probe measurement results indicate a matching better than −10 dB up to 315 GHz, the GGB probe results indicate a matching below −10 dB only up to 285 GHz. The minimum transmission coefficient S_{21}, shown in Fig. 4.17b is −3.2 dB and −5.3 dB at 330 GHz for the CM probe and GGB probe, respectively. Evaluating the S_{21} phase of the test structures in Fig. 4.17c shows, that the optimized bond-pad has no deviations from the expected behaviour. The overall CDR loss of the optimized bond-pad structure shows a maximum value of 41 % at 302 GHz, which is an improvement of 50 % compared to the PDK bond-pad, as can be seen in Fig. 4.17d. Generally, the improved performance of the optimized bond-pad is clearly visible in all characteristics of the test structure.

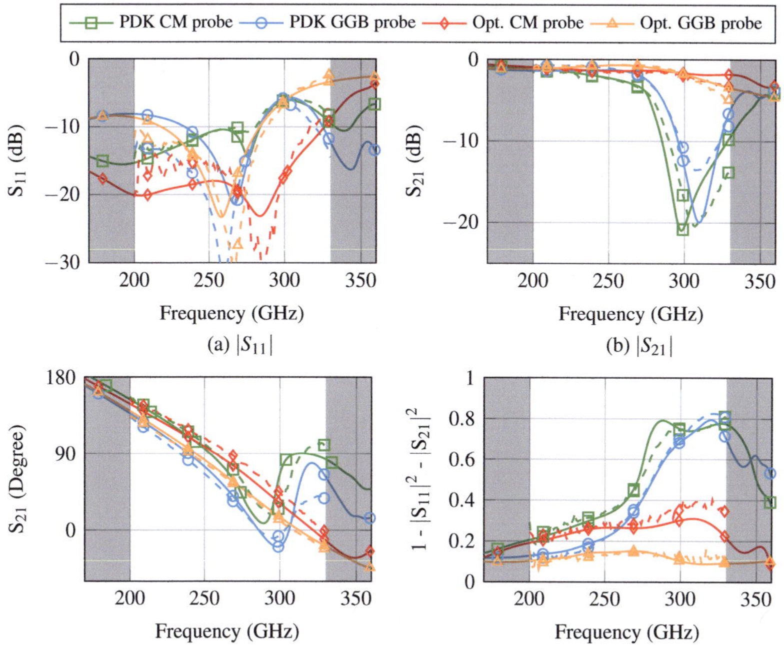

(a) $|S_{11}|$ (b) $|S_{21}|$ (c) $\angle S_{21}$ (d) CDR loss

Figure 4.17: Measurement (dashed) and simulation (solid) of the test structures for the PDK bond-pad and pad configuration D.

The good performance of the optimized bond-pad is also clearly visible by evaluating the E-field magnitude in Figs. 4.18 and 4.19, for the z-plane and y-plane, respectively.

To allow direct comparison, the cross-section planes and E-field magnitude range are identical to Figs. 4.7 and 4.8, which show the E-field magnitude

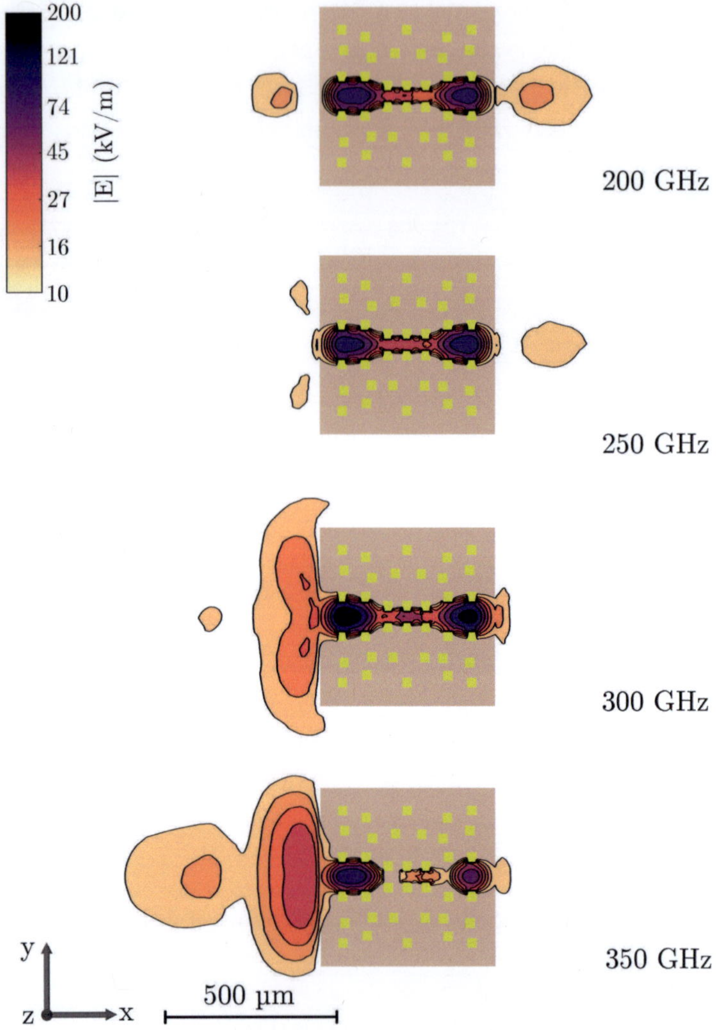

Figure 4.18: Z-plane cross-section at the center of the MMIC showing the E-field magnitude for different frequencies of the optimized bond pad test structure (feed left).

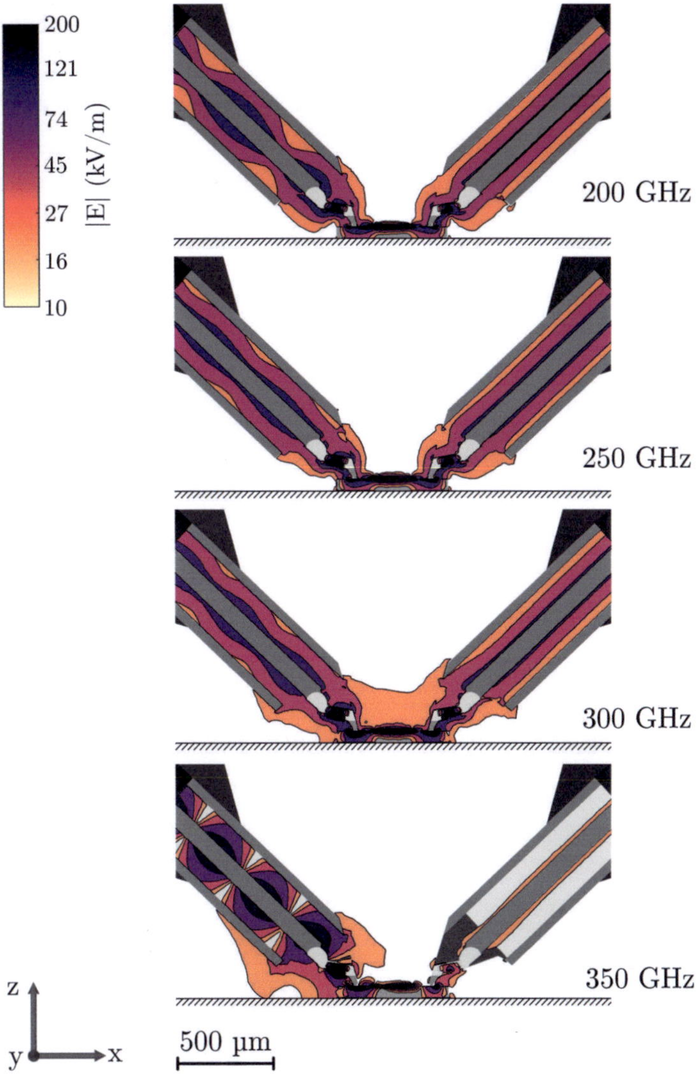

Figure 4.19: Y-plane cross-section at the center of the MMIC showing the E-field magnitude for different frequencies of the optimized bond pad test structure (feed left).

of the PDK bond-pad at the identical frequencies. The results indicate much less radiation compared to the PDK bond-pad.
Even at 350 GHz no parasitic modes are visible inside the substrate. Evaluating the y-plane cross-section shows basically the same field components between RF probe and chuck, but at a much lower magnitude, therefore having only minor influence on the circuit performance.

4.3.4 Pad resonance frequency depending on the EM model level of detail

Evaluating the measurement results of the unoptimized PDK bond-pad revealed a frequency offset of approximately 10 % between the analytically predicted resonance frequency and the measured resonance frequency. To investigate this difference, a simulative study for the PDK bond-pad was carried out.

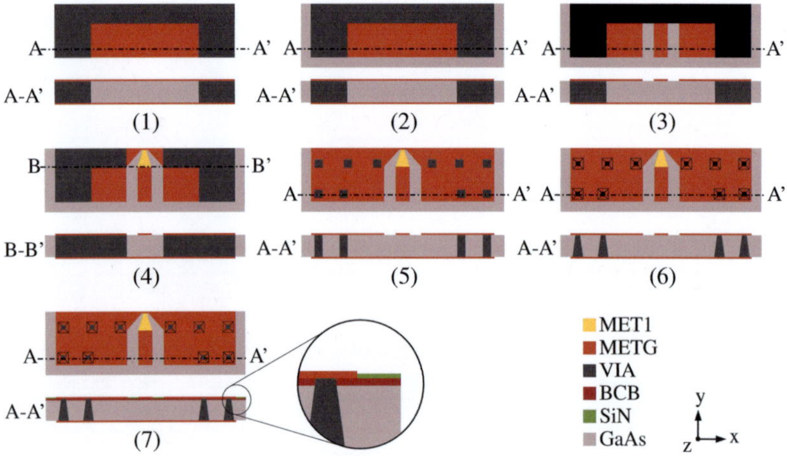

Figure 4.20: Different scenarios to investigate the transition from analytical model to practical realization of the PDK bond-pad. For clarification the layers BCB and SiN are not shown in the top views.

The model presented in section 4.2.2 is based on simplifications, which are necessary to allow the analytical investigation, but distort the accuracy of the predicted resonance frequency for practical pad structure realizations. Starting from the simplified model, the level of detail of the structure is increased step by step to finally meet the actual pad design. Schematic top and cross-section views of the different steps are shown in Fig. 4.20. The individual steps are:

1. Ideal, analytical assumptions, closed surface and vias represented by solid walls.
2. the ideal model is extended by surrounding substrate, necessary in practical realizations, which require a minimum distance between chip border and metal layers.
3. the closed surface is interchanged by the actual CPW pad structure.
4. the closed via wall is opened below the CPW transmission line taper to allow signal propagation.
5. ideal cuboid vias with a distance in accordance with the design rules replace the solid via walls.
6. vias approximating the real geometry modeled by truncated pyramids replace the cuboid vias.
7. technology dependent dielectric layers are added to the model.

Calculation of the corresponding resonance frequency for each step was done using the CST Eigenmode solver, which is intended to find eigenfrequencies in closed resonant structures [Tec17]. Except for z_{min} where the boundary was set to electric, due to the chuck, the remaining boundaries were chosen to magnetic. The simulation space was extended in the z_{max} and y_{min} direction with 100 µm vacuum to remodel the surrounding free space in an on-wafer measurement environment. The simulated resonance frequencies for each step are shown in Fig. 4.21 and are compared to the analytically calculated resonance frequency as well as the measured and RF probe tip simulated resonance frequencies. As can be seen, the analytically predicted resonance frequency and the one calculated by the eigenmode solver are in very good agreement for step 1.

The major shift in resonance frequency is due to the substrate extension of step 2, which expands the effective resonator volume. The subsequent steps further increase the resonance volume step by step, e.g. due to opening of the top metal layer to form a CPW transmission or due to vias, which allow electric field components between the vias. The inclusion of the dielectric layers in step 7 reduces the effective permittivity of the cavity and therefore increases the resonance frequency, resulting in a close match to the measured and RF probe tip EM simulated results of the resonance frequency of both RF probes. Evaluating the results shows that it is already possible with the reasonable level of detail in step 2 to give a very good estimation of the occurring resonance frequency.

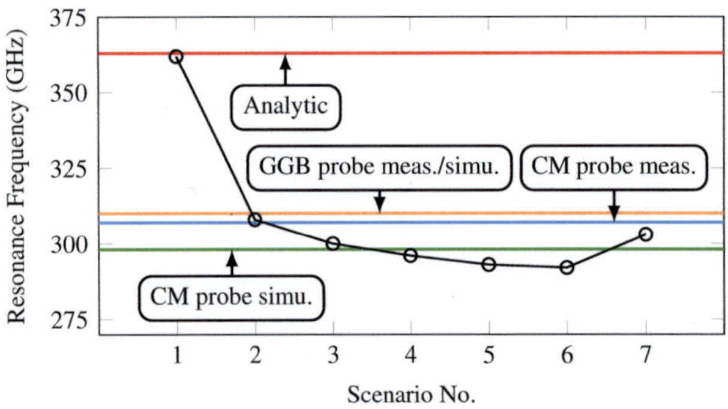

Figure 4.21: Resulting resonance frequencies for each step of the transition from ideal to practical pad structure, compared to analytical, measured and EM probe tip simulated results.

4.3.5 Sensitivity analysis of fabrication and measurement tolerances

Even though the EM simulations using the RF probe tip models provide a good prediction of the measured behaviour, certain deviations remain. To investigate these deviations and more generally the electrical robustness of the pad structures, the fabrication and measurement tolerances were investigated in detail. For these investigations the test structure of the PDK bond-pad and the optimized bond-pad were chosen. While the PDK bond-pad is prone to fabrication tolerances, due to its significant resonance in the frequency range of interest, the optimized bond-pad should provide a more robust behaviour. To have full control over the geometrical and electrical deviations, these investigations were carried out by means of EM simulations using the CM RF probe tip models. Furthermore, the results not only depend on the DUT tolerances but also on the error set of the calibration procedure. Therefore, the parameters were varied not only in the model of the DUT but also in the models of the calibration standards, which are used to calculate the TRL calibration error terms. For the DUT, several variables such as dielectric constant of the used materials, uncertainties in geometry and misalignment of the RF probes were taken into account. The dielectric constant of the substrate containing the calibration standards was varied and a RF probe placement uncertainty was investigated.
Since the dielectric constants of the GaAs substrate and dielectric passivation layers of the DUT are not precisely known at mmW frequencies [MRS01], an uncertainty of 5 % is assumed. Furthermore, a variation in conductivity of the metal layers of 10 % is investigated to model fabrication tolerances and surface roughness variation. Another fabrication tolerance is the size of the vias, which are realized with a backside etching process, resulting in a truncated pyramid like geometry with rather large deviations from an ideal truncated pyramid. A SEM picture of a via cross-section is shown in Fig. 4.22.

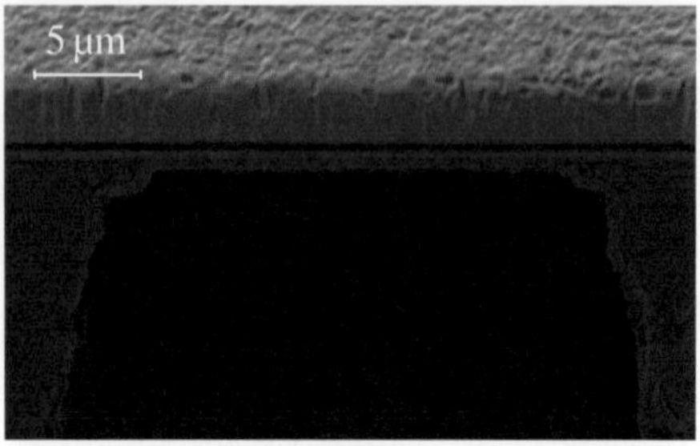

Figure 4.22: SEM picture of a via cross-section.

Again, a deviation of 10 % is assumed to represent the tolerances. This is equal to a size variation Δv of approximately ± 2.5 μm as shown in Fig. 4.23. The last fabrication tolerance which was examined on the DUT concerns the chip size after laser dicing of the MMICs. Due to the dicing process the resulting chip size is slightly smaller than in the layout, which was investigated by a decrease in chip size of 5 μm and 10 μm.
To investigate errors arising not by MMIC tolerances but due to the measurement setup, variations in RF probe placement on the DUTs contact pads were evaluated assuming an uncertainty in RF probe placement Δx and Δy of ± 5 μm in both lateral directions, as can be seen in Fig. 4.23. Although the pad has alignment marks included, this is due to the limited accuracy of the manual positioners itself and the limited magnification of the microscope a realistic uncertainty. The movement of both RF probes was assumed symmetrically in y-direction and opposed in x-direction, i.e. the probes were moving either inwards or outwards on both sides of the MMIC. Compared to the DUT, the geometry of the calibration substrate is much simpler and better controlled. Therefore, only the dielectric constant of the ISS substrate and the RF probe placement were varied. Similar to the DUT, the RF probe placement was varied by ± 5 μm in both directions. During the TRL calibration procedure the distance of the probes is adjusted for the thru and open standard with the help of alignment marks on the ISS and not changed

between the measurement of these two standards. Therefore, the offset for the thru and open standard was assumed identical. The line standard, however, has a different length and therefore a re-adjustment of the probe has to be performed, leading to an independent offset. The dielectric constant of the alumina (Al_2O_3) substrate was varied by ±5 %.

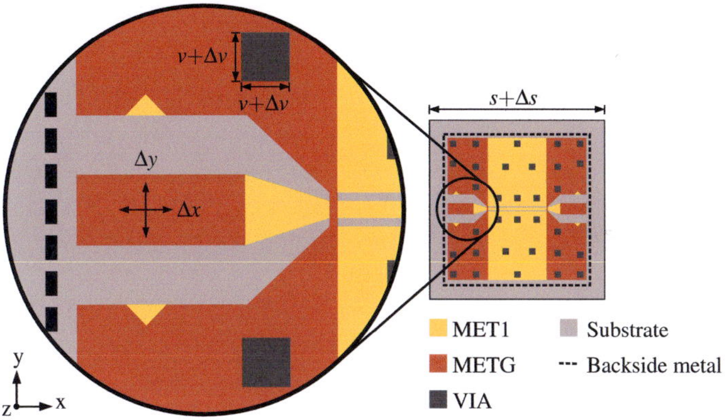

Figure 4.23: Detail view of the parameters investigated for modelling fabrication and measurement uncertainties.

Altogether, the following parameters were varied:

- $\varepsilon_{r,GaAs}$ of the DUT GaAs substrate by ±5%
- $\varepsilon_{r,BCB}$ and $\varepsilon_{r,SiN}$ of the DUT dielectric layers by ±5%
- Conductivity of the DUT metal layers by ±10%
- Size of the DUT vias Δv by ±2.5 µm
- DUT chip size Δs by −5 µm and −10 µm
- RF probe position on the DUT contact pad by ±5 µm
- ε_{r,Al_2O_3} of the ISS alumina substrate by ±5%
- RF probe position on the ISS calibration standards by ±5 µm

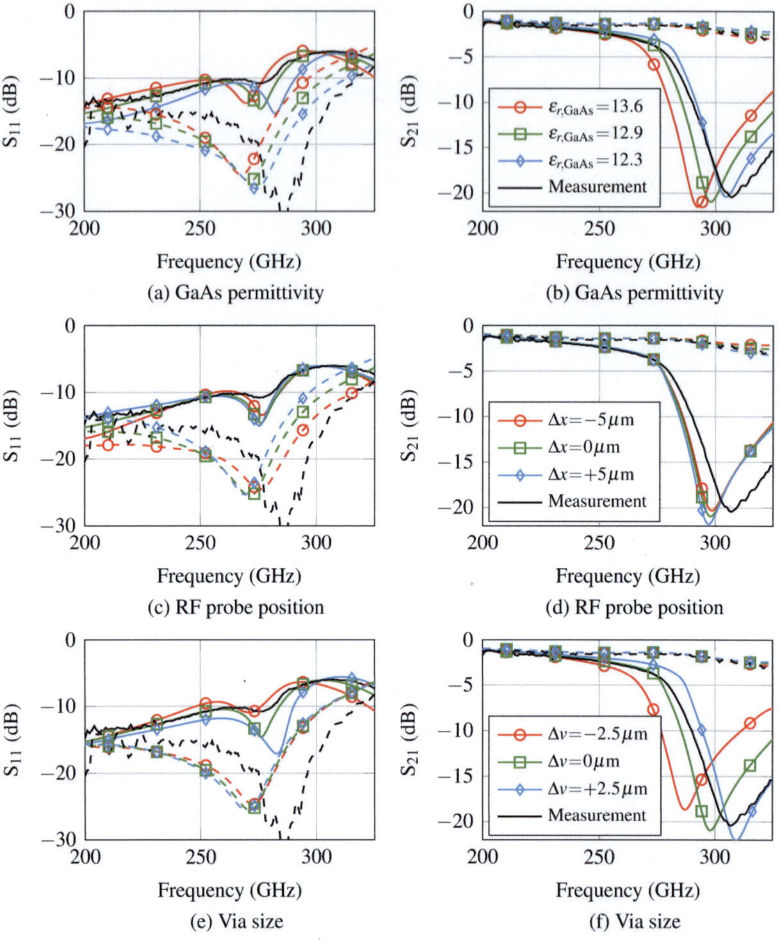

Figure 4.24: DUT fabrication and measurement tolerances evaluation for the PDK bond-pad (solid lines) and optimized bond-pad (dashed lines) test structures.

The results of the DUT parameter variations are summarized in Fig. 4.24. While the permittivity variations of the dielectric layers and conductivity variance of the metal layers showed only negligible influence, the change in permittivity of the GaAs substrate shows an impact on S_{11} and S_{21}. For the PDK bond-pad the change of $\varepsilon_{r,GaAs}$ by $\pm 5\%$ leads to a shift in resonance frequency of approximately $\pm 2\%$, as can be seen in Figs. 4.24a and 4.24b. Variance in the RF probe placement is shown in Figs. 4.24c and 4.24d. RF probe misalignment in y-direction showed only minor influence on the simulation results and therefore only the effect of x-direction misalignment is shown. Generally, the RF probe misalignment has only a minor effect on the resonance frequency, but results in deviations of S_{11}. Variation of the via size is evaluated in Figs. 4.24e and 4.24f. For larger vias the resonance frequency shifts to higher frequencies whereas for smaller vias the resonance frequency decreases. This is in good agreement to the analytical prediction, where the resonance volume is changed by increasing and decreasing the size of the vias. Finally, the variation Δs of the chip size showed no significant effect on the simulation results and is therefore not shown. Next, the influence of variations during the calibration procedure on the calibrated results was investigated and is shown in Fig. 4.25.
The resonance in S_{11} is sensitive to RF probe position and dielectric constant uncertainties, as can be seen in Figs. 4.25a and 4.25c. On the other hand, the simulations show that neither the change in substrate permittivity nor the RF probe position variation affects the calibrated transmission coefficient S_{21}, as can be seen in Figs. 4.25b and 4.25d.
To investigate the combined behaviour of the fabrication and measurement tolerances, permutations of all investigated parameter variations were calculated. This results in 736 curves, which cannot be plotted in a useful manner. Therefore, only the covered area of the resulting curves is shown in Figs. 4.26a and 4.26b for the input reflection coefficient S_{11} and transmission coefficient S_{21}, respectively. As can be seen, even small fabrication and measurement tolerances lead to large uncertainties in mmW measurements.

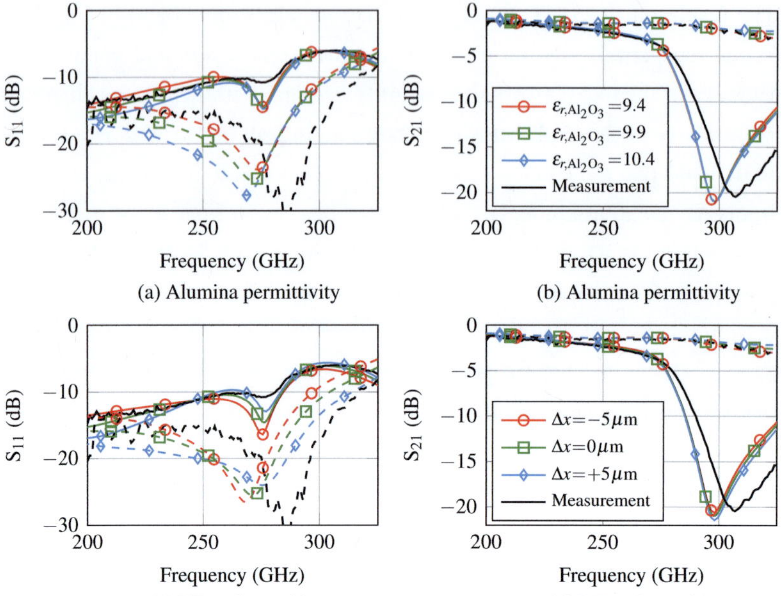

(a) Alumina permittivity

(b) Alumina permittivity

(c) RF probe position

(d) RF probe position

Figure 4.25: Influence of measurement and fabrication tolerances of the calibration standards on the PDK bond-pad (solid) and the optimized bond-pad (dashed) test structures.

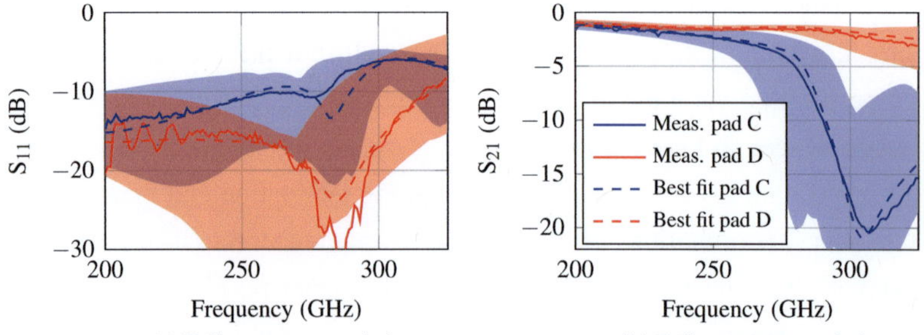

(a) Full parameter variation

(b) Full parameter variation

Figure 4.26: Permutations of all investigated fabrication and measurement tolerances compared to the measurement.

For frequencies well below the resonance the variation in transmission coefficient S_{21} is relatively low, while increasing in vicinity of the resonance frequency. The input reflection coefficient S_{11} on the other hand shows a strong dependency on the investigated variations, independent of the frequency range and proximity to the resonance.
The permutations were evaluated in order to find the best correlation with lowest RMS errors between simulation and measurement within the fabrication and measurement tolerances. The results shown in Figs. 4.26a and 4.26b support the assumption that the observed deviations between EM RF probe tip simulation and measurement result from fabrication and measurement tolerances.
To investigate which parameters have a major influence on the agreement between measurement and simulation the top 5 % of the varied curves with the lowest RMS errors were evaluated. This was done for the PDK bond-pad as well as for the optimized bond-pad and is shown as histograms in Fig. 4.27. The evaluation of the GaAs substrate permittivity variation for both bond-pad test structures in Fig. 4.27a shows, that a good agreement between measurement and simulation is achieved for slightly lower values than the usually used $\varepsilon_{r,\mathrm{GaAs}} = 12.94$. The probe position variation on the two bond-pad test structures is evaluated in Fig. 4.27b. While for the PDK bond-pad no distinct behaviour for a good fit is visible, the optimized bond-pad shows a tendency to probe positions shifted outwards of the MMIC. The evaluation of the via size extension, shown in Fig. 4.27c, shows a better match between measurement and simulation for unchanged or extended via sizes, in case of the PDK bond-pad. Evaluation of the optimized bond-pad on the other hand shows that the via size extension has no distinct influence on the agreement between measurement and simulation. The origin of this behaviour lies in the resonance of the PDK bond-pad, whose resonance frequency is significantly influenced by the size of the vias. Evaluating the Alumina substrate permittivity variation of the ISS calibration standards in Fig. 4.27d clearly shows that for slightly lower values a better fit is achieved.

The evaluation of the probe position variation for the open and thru standard, depicted in Fig. 4.27e, indicates that for positions directly at the intended position or shifted slightly outwards leads to good results. Opposed to this is the behaviour of the line standard, which results in a better fit for values directly at the contact position or shifted slightly inwards the MMIC, as can be seen in Fig. 4.27f.

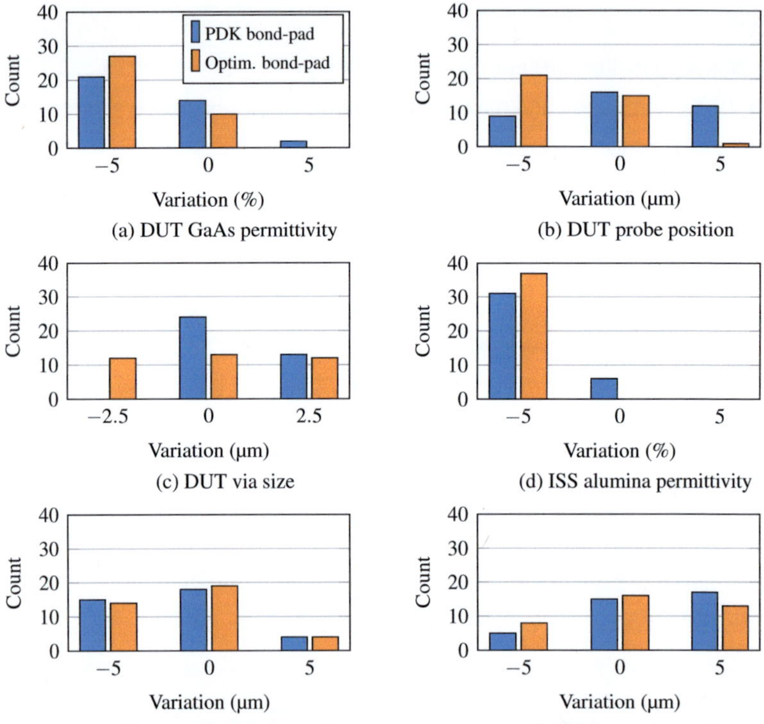

(a) DUT GaAs permittivity

(b) DUT probe position

(c) DUT via size

(d) ISS alumina permittivity

(e) ISS open/thru probe position

(f) ISS line probe position

Figure 4.27: Evaluation of the bond-pad test structures and ISS calibration standards variations for the best 5 % of curves with the lowest RMS errors to the measured behaviour.

4.3.6 Evaluation of the optimized bond-pad

To summarize the results of the different pad configurations, Table 4.1 lists the comparison of the resonance frequencies of the different test structures calculated analytically with (4.7), the EM simulated results and the resonance frequencies acquired by measurements, if available. The analytical method could not be applied to pad configuration A, which has no additional vias and therefore no defined confinement of the electric field.
The test structures for pad configuration A and B were not realized and therefore no measurement results are available. Generally, the influence of the additional vias in regard to the resonance frequency is clearly visible. For the PDK bond-pad (pad C), where results for analytical and simulated resonance estimation as well as the measurement are available, the relative error of the measured resonance frequency is 18 % to the analytic model and −2.9 % to the field simulation including the RF probe tips. The analytically calculated resonance frequency for pad configuration D is 531 GHz. However, extrapolating the shift of resonance frequency for pad configuration B and the PDK bond-pad, the resonance for test structure of the optimized bond-pad (pad D) is expected around 450 GHz. Furthermore, it has been shown that the observed deviations between EM simulation results including the RF probe tips and the measured deviations can be traced back to small fabrication and measurement inaccuracies.

Table 4.1: Resonance frequency (GHz) depending on pad configuration and calculation method compared to measurements.

	Pad configuration			
	A	B	C	D
Analytic (eq. (4.7))	–	337	363	531
RF probe simulation	233	270	298	>330
Measurement	–	–	307	>330

4.4 Optimized measurement-pads

Even though the optimized bond pad promises resonance-free operation in the WR3 frequency range, this pad is not an universal solution. Especially if the occupied chip area is of interest, which is the case in almost all designs, the required area of 0.046 mm^2 may be too large. Therefore, smaller pad geometries are investigated in the next section, which require a certain trade-off in terms of size and the variety of applications. The requirement for the pad structures was the suitability for a RF probe pitch of 100 μm.

4.4.1 Optimized standard measurement-pad

In analogy to the bond-pad optimization in section 4.3 the PDK measurement-pad was also optimized to suppress unwanted parasitic modes. The schematic layout of the optimized pad is shown in Fig. 4.28a and shows the dimensions along with the additional vias placed as close as possible to the inner conductor. The required area of the pad is 0.027 mm^2, which is only 60 % of the bond-pad area.

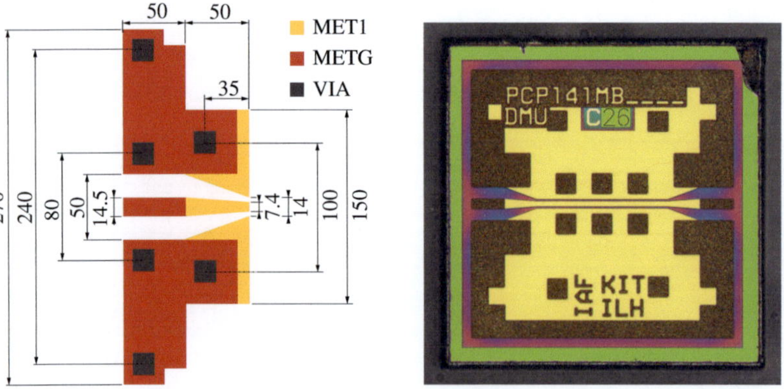

(a) (b)

Figure 4.28: (a) Sketch of the optimized measurement-pad (dimensions in μm) (b) microscopic photograph of the corresponding test structure. The chip size is $0.50 \times 0.50\,mm^2$.

A microscopic picture of the corresponding test structure is shown in Fig. 4.28b. The measurement results compared to ideal WG as well as RF probe tip EM simulation results are shown in Fig. 4.29. The input reflection coefficient of the test structure is shown in Fig. 4.29a and shows a strong dependency on the used RF probe. While the ideal WG simulation predicts an input reflection coefficient below -18 dB in the specified frequency range, the measured reflection coefficient S_{11} is only below -7 dB and -11 dB for the GGB probe and CM probe, respectively. The S_{21} transmission coefficient magnitude and phase, shown in Figs. 4.29b and 4.29c, show resonance-free behaviour of the test structure up to 330 GHz with a maximum loss of 4.2 dB and 2.3 dB for the CM probe and GGB probe, respectively. Above the specified frequency range the simulations using the CM probe show first signs of resonance behaviour while the GGB probe results stay close to the WG simulation results of S_{21}.
The higher transmission loss of the CM probe also affects the overall CDR loss of the test structure, as shown in Fig. 4.29d. Comparing the loss of both setups shows that measured and simulated loss, acquired with GGB Picoprobes, is even lower than predicted by the ideal WG simulation. This behaviour is attributed to crosstalk between the RF probe tips, which deteriorates the measurement and calibration accuracy.

4.4.2 Pad with backside metal recess

Besides adding additional vias for suppressing the unwanted microstrip mode another option is to create a local backside metal recess below the pad structure. Without the backside metallization the microstrip like mode cannot be excited due to the missing reference potential. The drawback of this approach is, that the MMICs using this pad must not be placed on a metal surface.

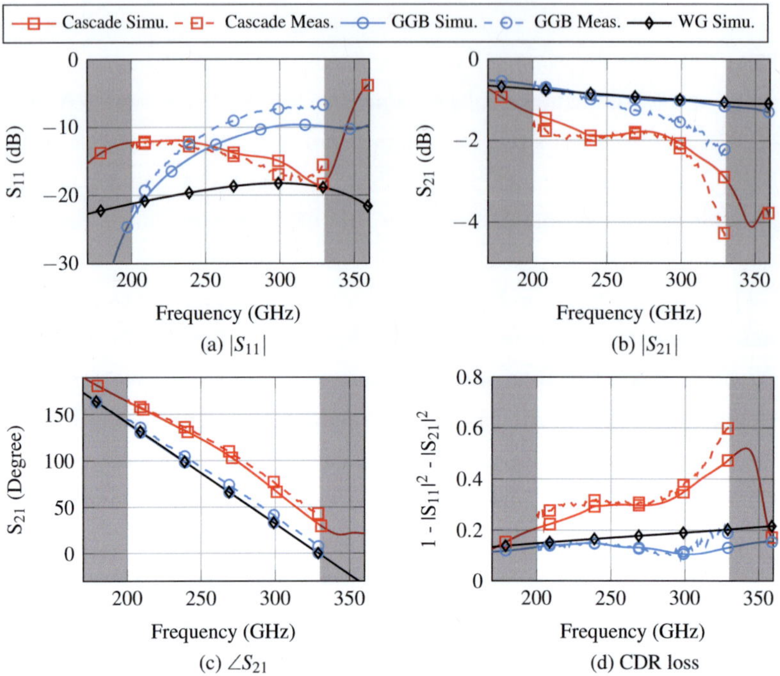

Figure 4.29: Measurement and simulation of the optimized measurement-pad test structure.

The underlying metal would act as ground plane for the parasitic microstrip mode i.e. for measurement purposes the MMICs have to be placed on a non-conductive material. Furthermore, for packaging purposes the environment below the pad has to be carefully considered and evaluated. A schematic drawing of the pad, including the dimensions, is shown in Fig. 4.30a. The dimensions of the inner conductor pad were chosen to be at the lower geometrical limit which can be handled by today's wire bond packaging technology. The RF probe tip landing area of the outer pins was chosen to allow a RF probe pitch up to 100 µm. Due to its backside metal recess, which is comparable to defected ground structures (DGSs) [KKK17], this pad is denoted as 'DGS-pad' in the following context.

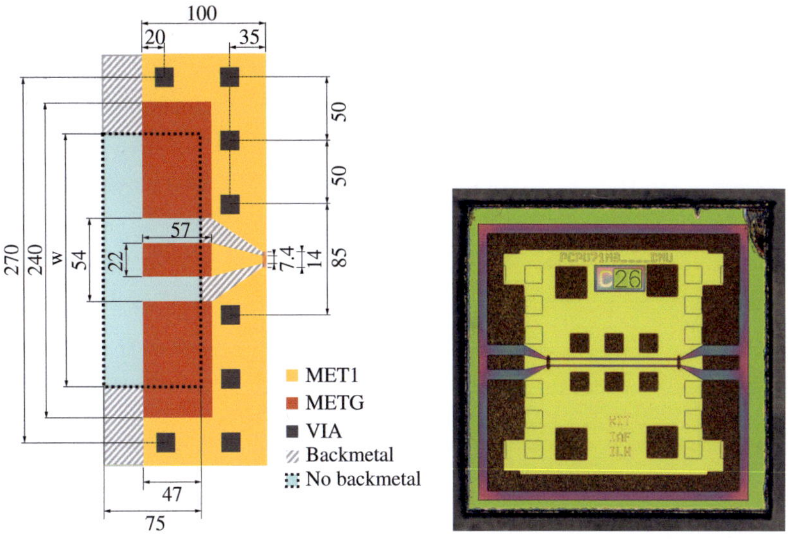

Figure 4.30: (a) Sketch of the optimized pad with localized backside metal recess (dimensions in µm) (b) microscopic photograph of the corresponding test structure.

To investigate the influence of the ground recess width *w*, depicted in Fig. 4.30a, EM models of test structures with different recess width *w* were created and simulated employing the CM RF probe tip model. In order to replicate the packaging environment, the boundary below the test structure was chosen to an infinite sheet of Rogers 6010, with a dielectric constant $\varepsilon_r = 10.2$. EM simulated results of the sweep were calibrated to the RF probe tip using the corresponding ISS error set. The impact of the width *w* on the reflection coefficient S_{11} is shown in Fig. 4.31a. While only a minor influence is visible at frequencies around 230 GHz and below, the effect of the recess width is clearly visible towards higher frequencies. Increasing the width efficiently decreases the reflection coefficient and a matching better than 15 dB is predicted over the full WR3 frequency range for widths larger than 150 µm. For recess widths above 200 µm no further improvement can be observed. In terms of the transmission coefficient S_{21}, it can be seen in Fig. 4.31b and 4.31c, that for a recess width w = 0, which equals a closed metallic backside, the simulation results are close to the PDK bond-pad and measurement-pad behaviour, with a strong resonance around 300 GHz. However, increasing the backside recess width shifts this resonance towards higher frequencies, making the pad suitable for frequencies up to 330 GHz. For widths above 150 µm no resonance can be observed in the WR3 frequency range any more. The CDR loss shown in Fig. 4.31d indicates that for a larger recess width w the loss of the test structure is decreasing, especially at the upper end of the WR3 band.

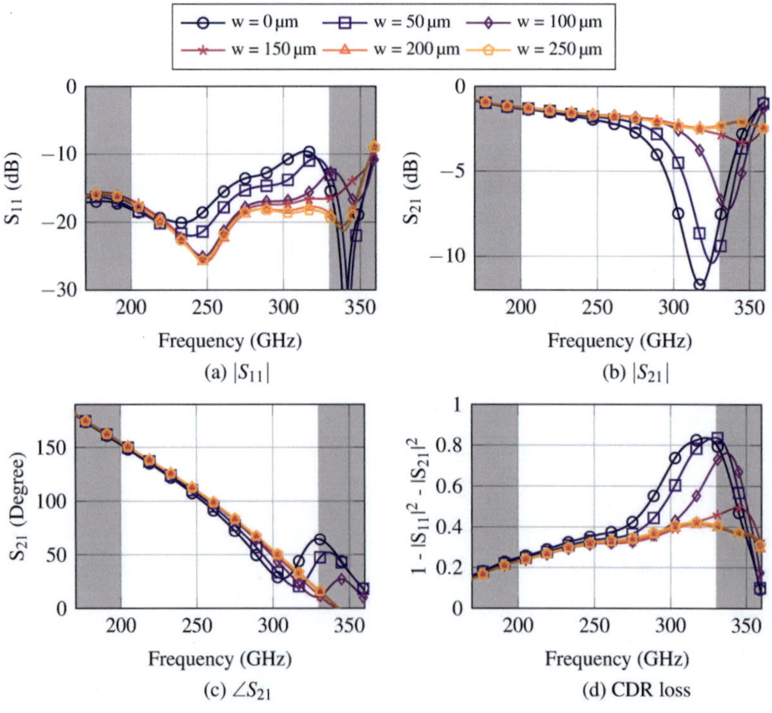

Figure 4.31: Simulated results of the ground recess width sweep for the DGS-pad using CM probes when placed on an infinite sheet of Rogers 6010.

To maximize the performance of the DGS-pad but on the other hand keep the recess width narrow to minimize the influence of the mounting substrate, the recess width was chosen to w = 220 µm and a corresponding test structure, which is shown in Fig. 4.30b, was designed and realized using the three layer Fraunhofer IAF 35 nm technology. This recess width is slightly higher than indicated by the simulations to include a back-up. The influence of the mounting substrate was investigated for three different materials with different electrical behaviour: conductive metal, Rogers 6010 with a thickness of 1.27 mm and PTFE block with a thickness of 5.3 mm. Placing the test structure on conductive metal is used to verify, if the microstrip mode is excited and how much this mode is distorting the high frequency pad performance. Rogers 6010 is a widely used substrate in mmW PCB designs and its dielectric properties with $\varepsilon_r = 10.2$ are close to GaAs, therefore acting as a substrate extension under the pad. PTFE with its low dielectric constant of $\varepsilon_r = 2.1$ was used to investigate an environment close to air, which is present if the MMIC is flip chip packaged. The simulated and measured S-parameter results on the different support materials are shown in Fig. 4.32. For frequencies below 300 GHz the different mounting materials show almost no influence on the behaviour of the test structure. Above 300 GHz there is a decrease in S_{21} for the test structure placed on conductive metal.

However, compared to the PDK bond-pad and PDK measurement-pad no resonance is visible in the WR3 frequency range, which is a result of a smaller resonator volume due to a more compact via placement. The test structures placed on Rogers 6010 and PTFE also show a decrease in S_{21} but with lower impact. The full verification of the DGS-pad test structure using GGB probes is carried out in appendix B.1. Generally, independent of the RF probe, the die placed on PTFE shows the lowest loss in the observed frequency range with a maximum loss of 2.3 dB and 2.9 dB observed at 330 GHz for the CM probe and GGB probe, respectively. The required chip area for the DGS-pad is 0.024 mm^2, which is only 52 % of the area required for the bond-pad.

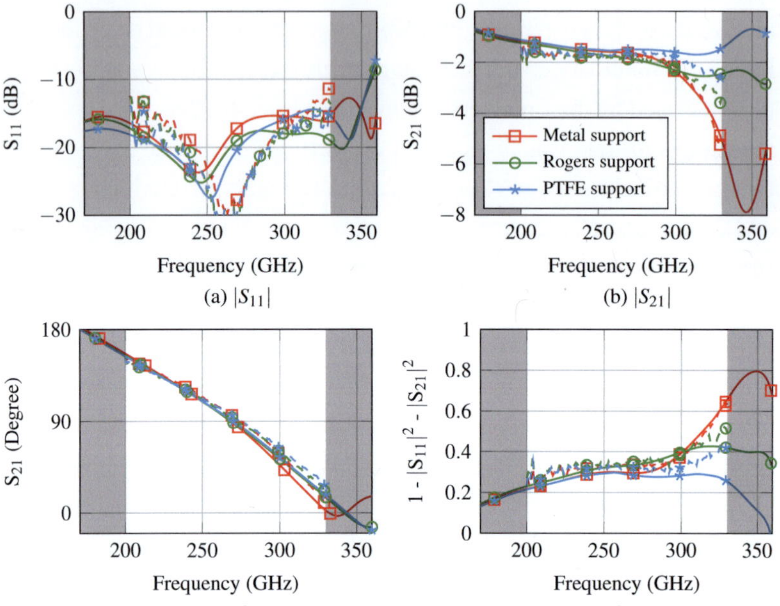

(a) $|S_{11}|$ (b) $|S_{21}|$ (c) $\angle S_{21}$ (d) CDR loss

Figure 4.32: Measurement (dashed) and simulation (solid) DGS-pad test structure contacted with CM probes.

4.4.3 Thin film microstrip pad

Another approach to suppress unwanted excitation in the substrate is to use a closed topmetal surface with no or only small holes. This allows only small field components to couple to the substrate and therefore robust operation over a wide frequency range is expected. This approach, however, requires at least two metal layers which are separated by a low-k dielectric since it is based on TFMS transmission lines. At the Fraunhofer IAF currently only the 35 nm technology supports TFMS transmission lines. One of the available metal layers is used for a closed metal surface and the second metal layer is used for transmission lines with BCB as a dielectric substrate. Due to this topology, the following pad is denoted as 'TFMS-pad' and the attached transmission line was chosen as a TFMS line. However, using a transition from TFMS to CPW, this pad can also be used for MMICs with a CPW transmission line environment, as will be shown in section 6.5. The TFMS-pad design uses a relatively small recess hole, in the otherwise closed MET1 surface, directly under the inner conductor pad area. This reduces the capacitance of the pad and allows the use of the full stack of all conductive process layers, which increases the mechanical stability due to direct contact on the GaAs substrate. The detailed dimensions of the pad as well as the arrangement of the different layers can be seen in Figs. 4.33a and 4.33b. A picture of the corresponding test structure is shown in Fig. 4.33c.

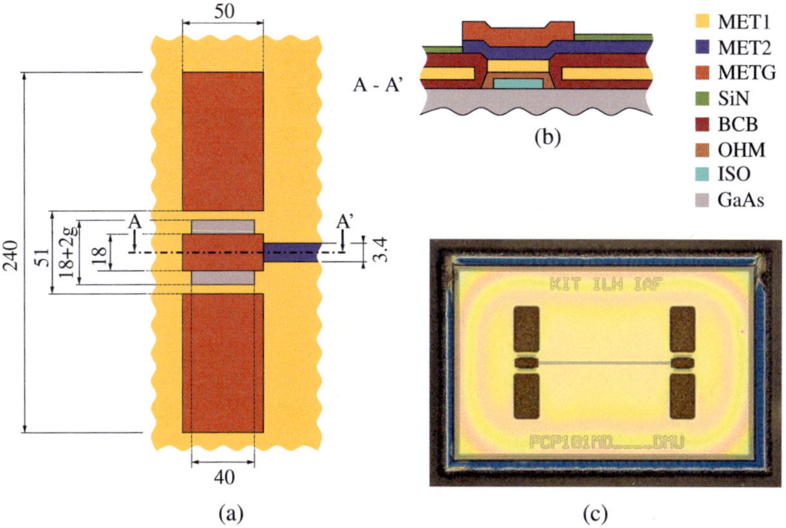

Figure 4.33: (a) Schematic top view (dimensions in µm) of the TFMS-pad design (SiN and BCB are not shown) (b) corresponding cross-section (c) chip photograph of the TFMS-pad test structure. The chip size is $0.75 \times 0.50\,\mathrm{mm}^2$.

The transmission line is realized using MET2 and forms a TFMS transmission line with the 1.6 µm thick BCB layer as dielectric and MET1 as ground. In the contact area of the pad, MET2 is extended over the recess length and enhanced with METG to form a large and mechanically robust landing area for the RF probe tip and for packaging. Similar to the DGS-pad, the width of the ground recess shows an influence on the overall pad performance. The length of the recess was chosen to 40 µm, which is a compromise between recess size and the mechanical robustness of the pad. The recess width on the other hand can be chosen more freely. Therefore, the gap g between inner conductor and MET1 was swept between 0 µm and 12 µm. The reflection coefficient S_{11} and transmission coefficient S_{21} results of investigation are shown in Fig. 4.34a and Fig. 4.34b-4.34c, respectively. The CM probe model deduced in section 3.1 was used for the simulations to precisely simulate the influence of the RF probe tips to the pad characteristics.

Evaluating the input reflection coefficient shows a direct correlation between gap width g and the resulting matching. For gap widths larger than 6 µm a matching better than −10 dB over the full WR3 frequency range is predicted. The same behaviour is observed for the evaluation of the insertion loss, where a larger gap results, due to the improved matching, in lower insertion loss. However, for gap widths above 6 µm there is only little improvement of insertion loss. The CDR loss of the corresponding test structures is shown in Fig. 4.34d. As can be seen, the CDR loss is almost independent of the gap width. In conclusion, to achieve good input matching with low transmission loss while keeping the gap as narrow as possible to prevent coupling to the substrate, the gap width of 9 µm was chosen.

In order to investigate the electrical robustness of the pad, no vias were added between top and backside of the realized MMIC. This however resulted in several resonances in the WR3 frequency range, as can be seen in the measurement and simulation results of the realized test structure shown in Fig. 4.35.

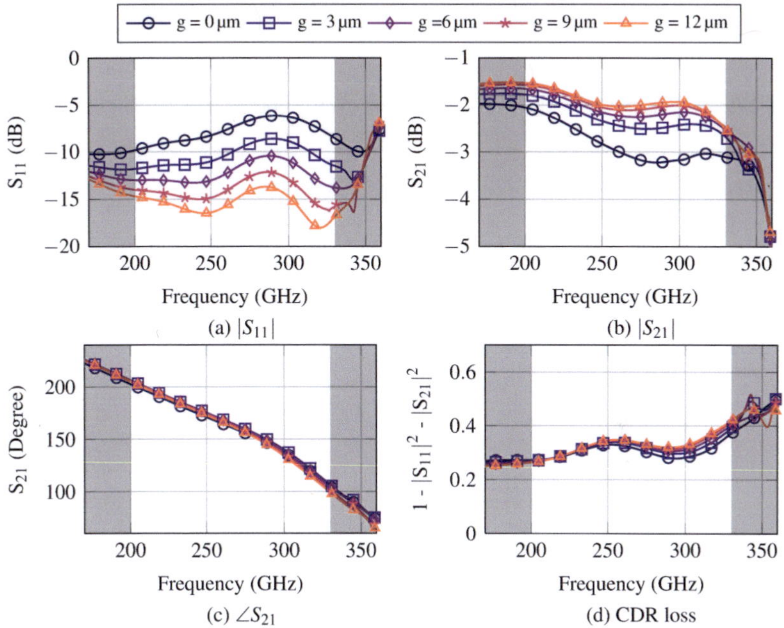

Figure 4.34: Simulated results of the ground recess width sweep for the TFMS-pad.

Evaluating the input reflection coefficient S_{11} in Fig. 4.35a shows generally a matching better than -10 dB but also shows first signs of the occuring resonances. In the transmission coefficient S_{21} the resonances are clearly visible throughout the whole WR3 frequency range, as shown in Fig. 4.35b. However, the distortions caused by these resonances are, except for the upper end of the specified frequency range, below 1 dB. To verify that these resonances are due to the absence of vias, a test structure with closely arranged vias with a distance of approximately 100 µm was EM simulated, including the CM probe tip models. Evaluating the simulations with additional vias clearly shows that these resonances can be easily suppressed. Due to the rather low distortions, the S_{21} phase characteristic shown in Fig. 4.35c has only negligible deviations from ideal behaviour. In terms of the CDR loss the pad test structure stays below 60 %, as shown in Fig. 4.35d. The analogue evaluation in case the GGB probe is used for simulation and measurements is carried out in appendix B.2. While the transmission coefficient S_{21} shows comparable behaviour to the CM probe measurements, the input reflection coefficient S_{11} degrades down to −6 dB.

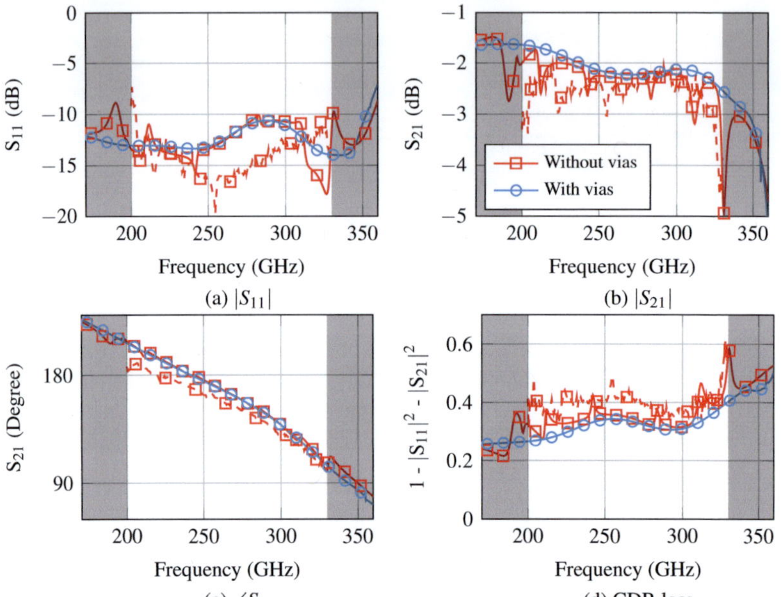

(a) $|S_{11}|$ (b) $|S_{21}|$ (c) $\angle S_{21}$ (d) CDR loss

Figure 4.35: Measurement (dashed) and simulation (solid) of the TFMS-pad test structure contacted with CM probes.

4.5 Conclusions on chapter 4

In this chapter, a comprehensive analysis on existing PDK pad structures as well as the optimization and design of on-wafer contact pads has been carried out. It has been found that the existing pad structures are prone to resonances and the excitation of parasitic modes around 300 GHz, which results in a high insertion loss and a degradation of the matching. These parasitic modes were investigated by means of an analytical model for the coarse prediction of the resonance frequency as well as simulations including the influence of the RF probe tips. It shows that, depending on the size of the via enclosed volume under the pad structure, strong resonances may occur, which deteriorate the overall circuit performance. Based on the EM simulation results, an optimized bond-pad and measurement-pad were developed, which provide resonance-free behaviour up to 330 GHz and presumably above. This was successfully verified by measurements using two different kinds of RF probes. Furthermore, the impact of fabrication and measurement tolerances on the pad performance was evaluated by means of EM simulations. It proves that the resonance frequency in S_{21} is insensitive to probe position variation but is directly linked to variations in the dielectric constant of the substrate or varying diameter of the vias. Furthermore, two additional pads were investigated, which also achieve resonance-free operation up to at least 330 GHz. These pads offer small size and low loss performance for MMICs, which are not intended for packaging purposes. Evaluating the loss per probe to pad interconnect of the different pads shows, depending on the used probe, values down to 0.9 dB for the DGS-pad and optimized measurement-pad. A comparison of all investigated pads is summarized in Table 4.2. To calculate the loss per interconnect, the loss of the connecting CPW (or TFMS) line is subtracted from the measured S_{21} and the result is halved. The results of this chapter can be used to optimize contact pads for even higher frequencies, which are required in sub-mmW and terahertz circuit design. Furthermore, it is possible to adapt the presented techniques to the design of contact pads for printed circuit boards (PCB) where comparable geometries are used.

Table 4.2: Comparison of size and probe to pad interconnect loss at 330 GHz for each investigated pad configuration and both RF probes.

Pad	Dimensions (μm^2)	Area (mm^2)	Loss per interconnect (dB) CM probe	Loss per interconnect (dB) GGB probe
PDK bond	400 × 115	0.046	9.75	6.75
PDK measurement	270 × 100	0.027	2.75	1.75
PDK characterization	270 × 337	0.091	0.85	0.6
Opt. bond	400 × 115	0.046	1.3	2.4
Opt. measurement	270 × 100	0.027	1.9	0.9
DGS	240 × 100	0.024	0.9	1.2
TFMS	240 × 50	0.012	0.55*	0.58*

* simulated

5 Evaluation and optimization of calibration standards

As discussed in the previous chapter, the interaction between RF probe and the contact pad can have a severe influence on the overall DUT performance. However, if the DUT is not intended for packaging but only for characterization of the internal circuit, it is a viable option to shift the calibration reference planes behind the actual probe to pad interconnect. This technique requires on-wafer calibration standards using the same substrate and contact pad as the DUT. The influence of the contact pad structure can then be included in the calibration error set and corrected in the measurement results. The basic idea of this concept is shown in Fig. 5.1 and was applied to the Fraunhofer IAF process in [See+15]. Since the probe to pad interconnect is embedded in the error terms, the probe tips do not have to be included into the EM simulation, resulting in much lower simulation volume and therefore shorter simulation times. However, to achieve a high accuracy on-wafer calibration, the corresponding calibration structures have to be carefully designed and need to fulfil several criteria:
As already mentioned, the calibration standards have to use the same contact pad structure and the same substrate as the DUT. Furthermore, the environment has to be identical in both MMICs e.g. adjacent structures or asymmetries have to be avoided or placed at a distance where the influence on the DUT or calibration standards can be neglected. Additionally, it has to be ensured that the overall calibration structure is not prone to the excitation of parasitic resonances, which will distort the accuracy of the error terms and therefore the measurement results. For example, during measurement of the

thru and line standards, energy which is coupled into a resonance will be falsely accounted as transmission line loss and therefore over-compensated after the calibration is applied to an actual measurement of a DUT.

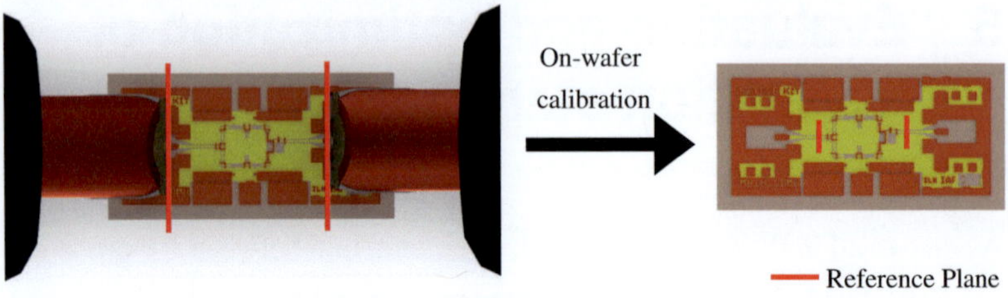

Figure 5.1: Reference planes of ISS and on-wafer calibration. For the on-wafer calibration, the reference planes are behind the probe to pad interconnect, allowing to neglect the RF probe in the simulation.

5.1 Analysis of existing on-wafer calibration standards

The existing on-wafer calibration structures from the PDK make use of the characterization-pad, which was presented in [See+15] and discussed in section 4.1. Fig. 5.2 shows a photograph of the PDK calibration structure with the reference planes outlined. The standards comprise two transmission lines of different length as well as short, open and a 50 Ω match. This variety of standards allows different calibration procedures such as SOLT, LRRM and TRL. As discussed in section 1.2.2, SOLT and LRRM require at least partial knowledge of the used standards, which is challenging at mmW frequencies and therefore the TRL calibration is utilized. The length difference of both transmission lines is 100 µm, therefore allowing TRL calibrations in the frequency range from 71.6 GHz to 576 GHz. After calibration, the reference plane is 150 µm behind the pad and the remaining thru and line standards are equal to 70 µm and 170 µm long CPW transmission lines, respectively. The calibrated open, short and match standards are connected with a 85 µm long transmission line to the reference plane, which is equal to an offset delay of 0.65 ps. In addition to the calibration standards, different transistor configurations for parameter extraction are also included on the MMIC but are not utilized in this work.

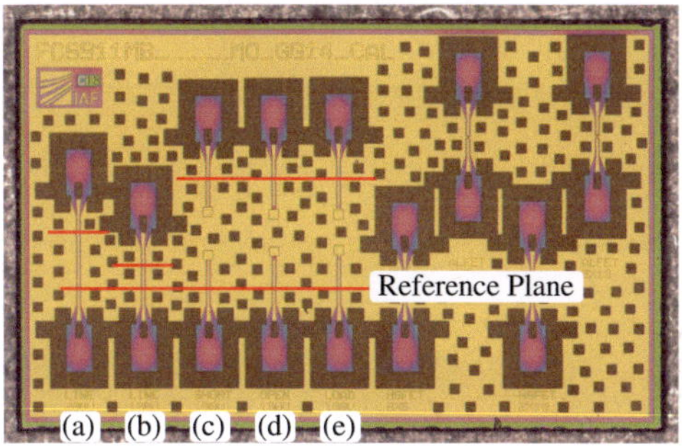

Figure 5.2: Photograph of the on-wafer calibration standards using the characterization-pads (a) line (b) thru (c) short (d) open (e) match. The reference planes 150 µm behind the pad are marked by red lines.

For the following investigations, measurements of the standards thru, line and short were used to calculate a TRL error set. Additionally, the standards were EM simulated using CST MWS in two configurations. The first configuration consisted of the full on-wafer calibration structure and the different standards were contacted using the CM probe as well as the GGB probe models presented in section 3.1. These results are used to calculate simulated TRL error terms. For the second EM simulation setup, WG ports were placed internally in the CPW transmission line at the reference planes of the various standards. For an ideal on-wafer calibration this should be equal to the TRL corrected simulation and measurement results. Fig. 5.3 shows both simulation setups exemplarily for the CM probe while contacted on the thru standard. The on-wafer calibrated measurement and simulation results of the various standards are summarized in Fig. 5.4.

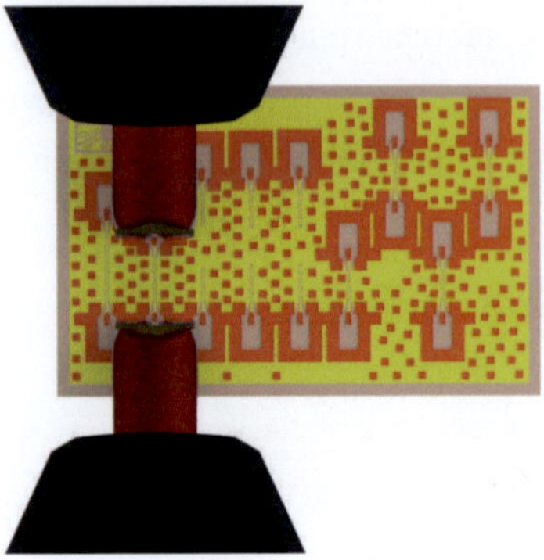

(a)

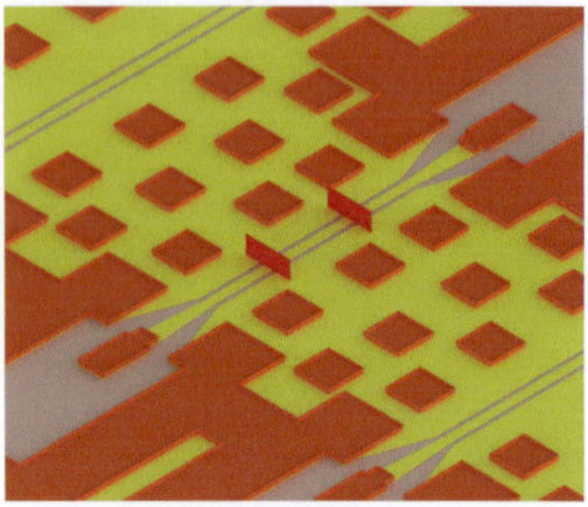

(b)

Figure 5.3: Simulation setup for (a) simulating the whole structure including the probe tips and (b) simulating the standard between the reference planes.

The results of the open and short standard are shown in Figs. 5.4a and 5.4b, respectively. For clarity, the corresponding sections of the Smith diagram are shown as cut-out.

Due to the offset delay of 0.65 ps the resulting S-parameters are rotated by approximately 140°, i.e. the short appears close to the open region of the Smith diagram and vice versa. While the expected behaviour of an offset reflect is verified by the WG simulation, the measurements and EM simulations including the probe tip models show several distortions in the specified frequency range from 200 to 330 GHz. For the CM probe measurement and simulation results these deviations lead to a reflection coefficient S_{11} of the open standard larger than unity, which is unphysical for passive devices.

For both line standards a monotonic decrease in transmission magnitude towards higher frequencies was expected and is also predicted by the EM simulation with internal WG ports. Figs. 5.4c and e show that this is approximately fulfilled up to 250 GHz for measurement and probe tip EM simulation results.

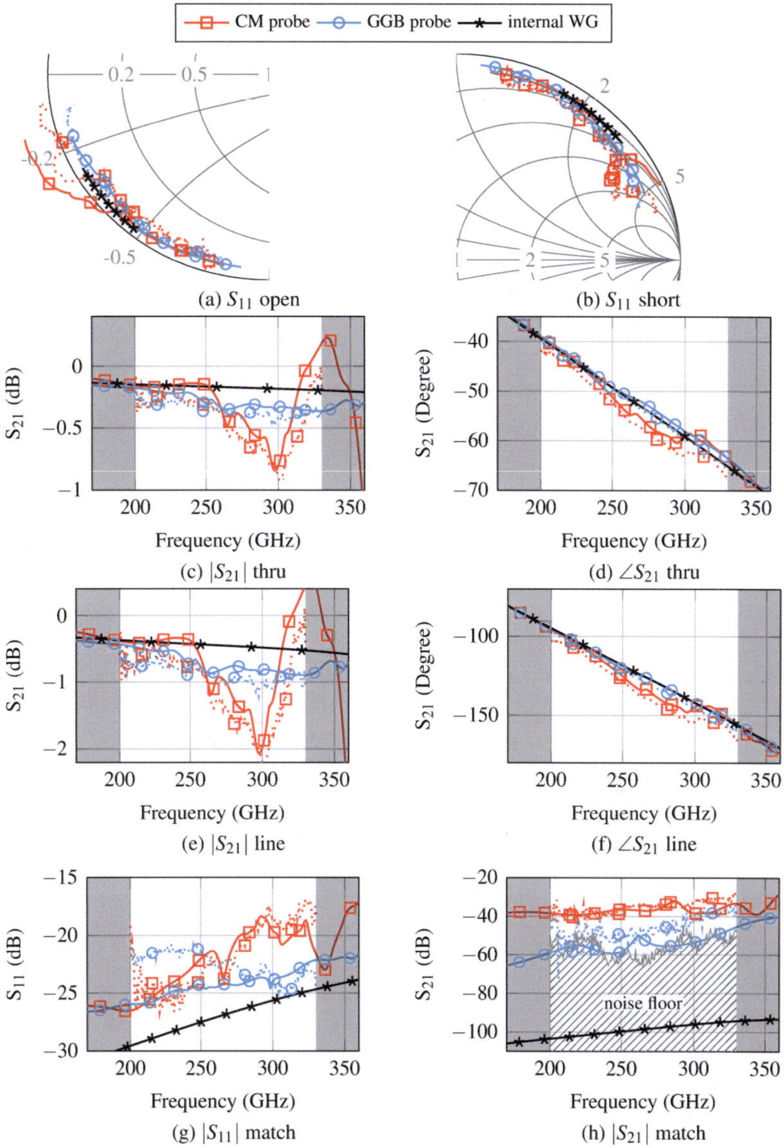

Figure 5.4: Comparison of on-wafer calibration corrected measurements (dotted) and simulations (solid) of the PDK on-wafer calibration standards.

While the results of the GGB probe shows only little deviations towards higher frequencies, the results of the CM probe show two major distortions at 263 GHz and 295 GHz. Towards the upper band edge the calibrated transmission coefficient S_{21} of the thru and line standards show measured values close to zero, whereas the probe tip simulation even predicts gain around 330 GHz, which is again unphysical. The transmission phase of the line standards is shown in Figs. 5.4d and 5.4f and shows that the observed distortions do not only affect the magnitude but also the resulting phase.
For further verification, the match standard, which was not used in the calibration procedure, was evaluated and the results are shown in Figs. 5.4g and 5.4h for the reflection and transmission coefficient, respectively. The reflection coefficient of both probes shows deviations from the WG simulated behavior with a decrease in matching and distortions at the previously observed resonance frequencies of 263 GHz and 295 GHz. The transmission magnitude shows a coupling up to -30 dB. Since the coupling using the WG excitation is below -80 dB, this high transmission coefficient is accounted to direct coupling between the probe tips. Generally, the measurements and simulations using the GGB probe show lower distortions in magnitude and phase for all evaluated standards compared to the CM probe. Especially the measurement and simulation results of the CM probe, which result in positive magnitudes are a strong indication of unwanted parasitic modes in the calibration standards.
The good agreement of the S-parameters between measurements and EM simulations including the probe tips allowed for investigation of the E-field distribution in detail and therefore investigate the origin of the distortions of both used probes. Fig. 5.5 shows the E-field distribution inside the substrate of the calibration structure when the thru standard is contacted. The fields are evaluated at 200 GHz and 330 GHz, which are the lower and upper end of the specified frequency range, and at 263 GHz, which is at the first resonance frequency of the line standards.

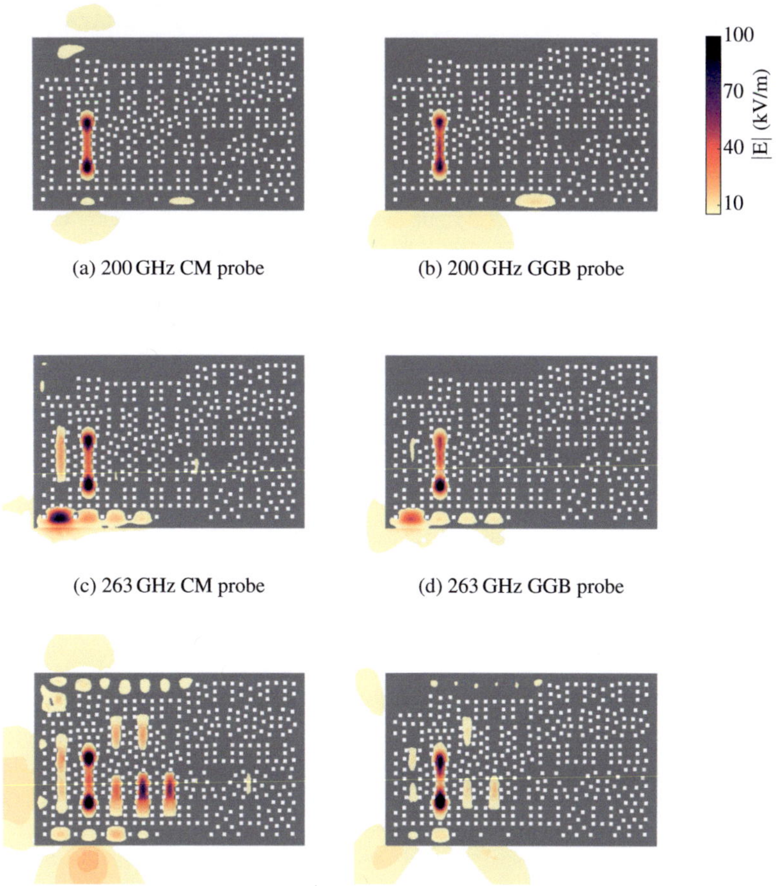

(a) 200 GHz CM probe (b) 200 GHz GGB probe

(c) 263 GHz CM probe (d) 263 GHz GGB probe

(e) 330 GHz CM probe (f) 330 GHz GGB probe

Figure 5.5: Field distribution inside the substrate of the existing on-wafer calibration structure at different frequencies and for different probes.

As shown in Figs. 5.5a and 5.5b the E-field at 200 GHz is mainly concentrated along the transmission line for both probes, therefore fulfilling the expected behaviour. Towards higher frequencies several modes at the border of the chip are excited and coupling to adjacent calibration standards occurs, as shown in Figs. 5.5c and 5.5d. As can be seen, the GGB probe excites the same modes as the CM probe but the coupled energy is lower by a linear factor of two. The EM field distribution at 330 GHz is shown in Figs. 5.5e and 5.5f where various resonances and strong coupling to adjacent standards can be observed. Again, the energy coupled to parasitic modes and adjacent structures is lower in case of the GGB probe. Generally, the CM probe couples more energy into parasitic modes, which radiate at the edges of the MMIC and below the RF probe. Another reason for distortions is especially visible at 330 GHz, where a significant amount of E-field is coupled into adjacent calibration standards. In the calibration algorithm this parasitic coupling and radiation is mistakenly accounted for as transmission line loss and therefore the accuracy of the calibration degrades. Further deterioration results from the asymmetry of the geometrical realization of the calibration standards which results in unequal S-parameter matching coefficients.

5.2 Design of optimized on-wafer calibration standards

The previous section discussed the characteristics and the parasitic distortions of the existing calibration structure. Using the probe tip EM simulations, it was possible to comprehend the origin of the deviations, which allows the development of improved on-wafer calibration standards. The new calibration standards were specifically designed for TRL calibration, therefore only three standards were necessary, which saves chip area and cost. The length difference of the transmission lines was chosen to 140 µm, which meets the TRL length difference criteria in the frequency range from 52.6 to 410.8 GHz. Below 52.6 GHz the spatial extend of the contact pad to the reference plane is below 1/40th of a free-space wavelength and therefore an ISS calibration with reference planes at the probe tip can be used with only minor influence to the measurement results. The reflect standard was realized with a shorted CPW transmission line, which represents a more defined reflection compared to an open end. Furthermore, the optimized calibration standards were designed fully symmetrical to avoid any asymmetric distortions of the standards. The influence of different on-chip environments surrounding the calibration standards was investigated as well as the influence of the geometrical length and spacing of the standards. The following investigations were carried out using the CM probe model, since this type of probe showed higher parasitic effects compared to the GGB probe. It is assumed that a substrate designed to work properly with CM probes, will also work with GGB probes without parasitic effects.

5.2.1 On-wafer calibration standard environment

After the basic calibration standards were defined, different environments shown in Fig. 5.6 were investigated. Common to all topologies is the number and placement of vias directly surrounding the standards, which are intended to electromagnetically shield the standards from the substrate. The structure shown in Fig. 5.6a consists only of the standards with minimal extent of the surrounding metal, which should avoid any parasitic parallel plate modes. In Fig. 5.6b a closed layer of metal is used to surround the calibration standards. This is intended to shield the substrate against external coupling. For the last environment shown in Fig. 5.6c, additional vias were added to the structure, which are intended to suppress any excited substrate modes. All structures were EM simulated including RF probe tip models and for each environment a corresponding TRL error set was calculated based on the simulation results. The reference planes were chosen in compatibility to the existing on-wafer calibration standards 150 µm after the pad, resulting

in line lengths of 0 µm and 140 µm for the thru and line standard, respectively. The short standard is directly located at the reference plane and has therefore no offset delay.
Summarized, the workflow is as follows:

- Design and layout of the on-wafer calibration standards
- Conversion into CST MWS according to section 2.1
- Simulation of the thru, line and short standard with RF probe models
- Calculating TRL error set based on the simulations
- Correction of the simulation results by the TRL error set

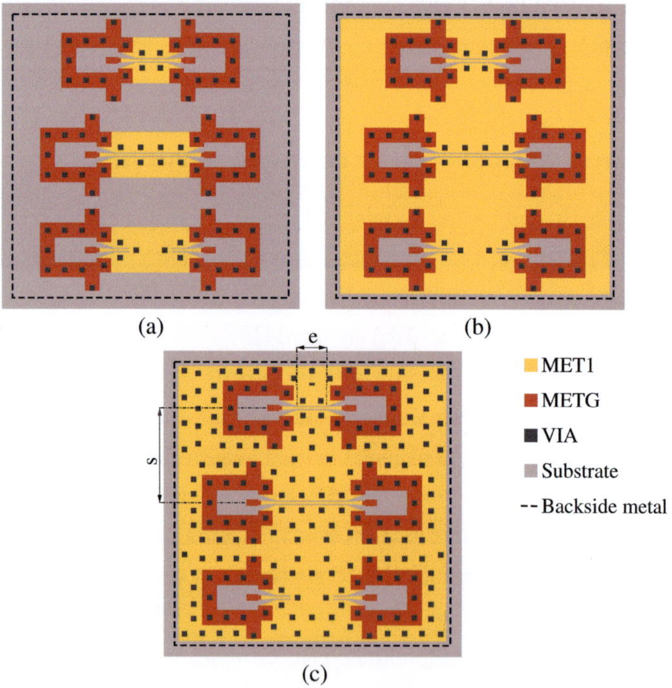

Figure 5.6: Investigated variations of the optimized on-wafer calibration standards (a) no surrounding top metal (b) surrounding top metal (c) surrounding top metal with additional vias.

In addition to the on-wafer calibrated EM simulation results using the probe tips, the standards were also simulated using internal WG ports. For the line standard simulation the WG port was placed directly at the reference plane. However, using the same location for the short standard would result in zero distance between port and the actual short location. Since this would result in inaccurate EM simulation the WG port was shifted back by 25 µm and then de-embedded by the same distance.

The results of the on-wafer calibrated standards for the different environments are shown in Fig. 5.7. The calibrated thru standard is not shown due to the fact that after calibration this appears as a perfectly matched line with zero length and zero loss. The simulation results using the internal waveguide ports do not depend on the environment of the standards, therefore only one internal WG simulation is shown. Evaluation of the transmission coefficient S_{21} in Fig. 5.7a shows that without the closed metal surface small resonances in S_{21} of the line standard are visible in the frequency range up to 250 GHz. Towards higher frequencies the distortions increase with a maximum loss at 300 GHz.

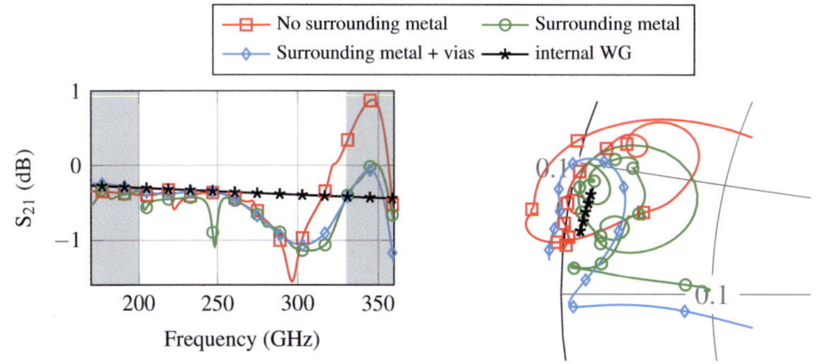

(a) $|S_{21}|$ line (b) S_{11} short

Figure 5.7: EM simulated influence of different environments on the characteristics of the optimized on-wafer calibration standards.

The simulation results of the closed surface show that without the additional vias several parasitic resonances occur in the WR3 frequency range, again with a maximum loss around 300 GHz but not as severe. The use of additional vias results in a smooth behaviour, which is very close to the ideal result of the internal WG up to 250 GHz. Above 250 GHz the EM probe tip simulated line standard also shows an increasing loss with a maximum around 300 GHz.

The peak loss around 300 GHz is common to all investigated environments. The origin of this will be investigated in more detail in section 5.2.2 and 5.2.3. Analysing the results of the reflection coefficient S_{11} of the short standard shows that without additional vias several resonances occur in the WR3 frequency range. The use of vias partially suppresses these resonances and smooths the response of S_{11}, as can be seen in Fig. 5.7b. But even with additional vias, the short standard shows a slightly positive reflection coefficient up to 240 GHz, which is unphysical and an indicator of small remaining parasitic effects of the calibration standards. Generally, the use of additional vias results in the most promising behaviour of the different environments and was therefore chosen for the following investigations.

5.2.2 Influence of spacing and length of the standards

Following the investigations on the surrounding environment, the influence of the geometrical length and spacing of the standards was evaluated. The geometrical parameters are shown in Fig. 5.6 and involve the length extension e of the standards and the spacing s between adjacent standards. Two parameter sweeps, one with a fixed length $e = 100\,\mu\text{m}$ and one with a fixed spacing $s = 310\,\mu\text{m}$ were performed. The values of the parameter sweeps are given in Table 5.1. For the investigations, the calibration structures with different spacing s and length e were layouted and converted into the EM simulation. The three standards thru, line and short were excited using the

CM probe tip model and the simulation results were first used to calculate a TRL error set and then to correct the corresponding simulations.

Table 5.1: Investigated length extension *e* and spacing *s* parameter variations.

Set 1	
e (μm)	*s* (μm)
100	200
100	310
100	410

Set 2	
e (μm)	*s* (μm)
100	310
200	310
350	310
600	310

The calibrated simulation results of the line and short standard for varying spacing *s* are shown in Fig. 5.8. Generally, in the specified frequency range the behaviour of the transmission coefficient S_{21} of the line standard is not significantly influenced by the spacing, as shown in Fig. 5.8a. Up to 250 GHz, the results follow the internal WG simulation closely. Since the peak loss around 300 GHz shows only minor dependency on the spacing between the standards, it is concluded that this behaviour is not caused by coupling to adjacent standards. The calibrated simulation results of the short standard are shown in Fig. 5.8b and also show only little influence of the spacing *s*.

Since the calibration structure shows only little dependency on the spacing of the standards, the next step was to verify the dependency of the length of the standards. To do so, the transmission lines and the reflect standard were extended to various lengths, which are given in Table 5.1. The length of the line standard was kept 140 μm longer than the thru standard and the reflect standard was kept at half the length of the corresponding thru standard. Using the simulation results, corresponding TRL calibrations were calculated for each sweep step with the reference plane at the center of the thru standard.

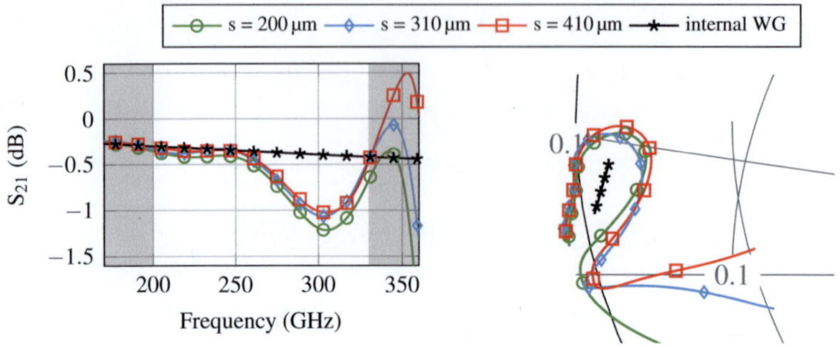

(a) $|S_{21}|$ Line (b) S_{11} Short

Figure 5.8: EM simulated comparison of different spacings between the standards of the optimized calibration structure (a) line (b) short.

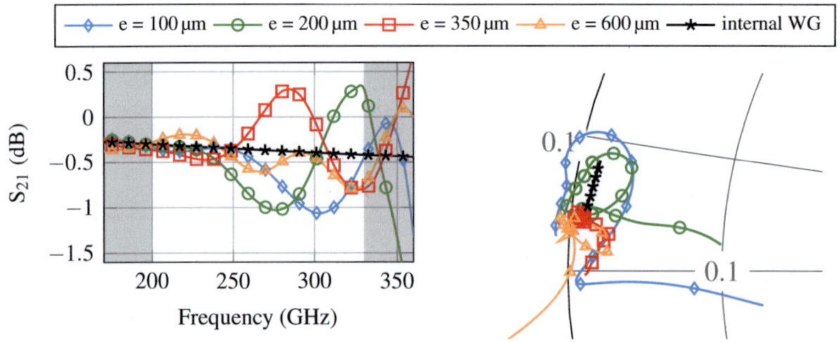

(a) $|S_{21}|$ line (b) S_{11} short

Figure 5.9: EM simulated comparison of different lengths of the standards of the optimized calibration structure.

The calibrated simulation results are shown in Fig. 5.9. When increasing the length of the standards, the local minimum in S_{21}, which until now was observed around 300 GHz moves to lower frequencies and an overlaying oscillation appears, as shown in Fig. 5.9a. Towards the maximum investigated extension $e = 600\,\mu\text{m}$, the magnitude of the overlaying oscillation decreases, which indicates that the distortions depend on coupling between the probes either directly over the air or through the substrate. Similar effects can be observed in the reflection coefficient of the short standard shown in Fig. 5.9b. Towards higher extensions e the area in the Smith chart concentrates, which results in a better fit to the internal WG simulation. However, despite the extension of the standard lengths up to $e = 600\,\mu\text{m}$, which besides would require an extensive amount of chip area, neither the line nor the short standard are in good agreement to the internal waveguide simulation. This diminishes the use of this approach.

5.2.3 Crosstalk analysis of optimized calibration standards

As presented in the previous section neither increasing the vertical spacing nor increasing the length of the calibration standards results in a good agreement to the expected characteristics. In order to investigate the origin of the deviations, additional simulations were carried out using the spacing $s = 310\,\mu\text{m}$ and extension $e = 100\,\mu\text{m}$. The decreasing deviations towards larger extension e indicate that the deviations are caused by RF probe crosstalk. Therefore, different scenarios were simulated to suppress this crosstalk. In the first simulation setup a thin sheet of absorbing material was inserted in-between the two RF probe models to minimize the direct crosstalk. As can be seen in the simulation results shown in Fig. 5.10, this approach lowers the deviations from the results obtained by the internal WG simulations. However, due to geometrical constraints the absorber is not able to fully suppress the direct probe crosstalk. Furthermore, due to the vicinity of the absorptive sheet to the RF probes, a direct interaction is observable, which will distort the calibration accuracy if the absorber is removed and an actual DUT is contacted. For further verification, the EM simulation model of the calibration standards were modified. A WG port was added at the center of the structure, i.e. two simulations were performed one for the left part and one for the right part of the calibration standard. After the simulation was done the two individual simulations were concatenated and calibrated (denoted as split + merge). Since the WG port placed at the center does only consider energy present in a CPW mode, this approach does not take into account any other propagating fields. As shown in Fig. 5.10, the transmission coefficient of the line standard now results in a very close match to the internal WG simulation. The reflection coefficient of the short standard still shows deviations from the expected behaviour, but compared to the previous simulations the best match is achieved. This indicates that the distortions are caused by parasitic coupling between the RF probes. However, it is not possible to suppress the crosstalk successfully in a real world measurement situation and therefore other techniques for correcting these errors are required.

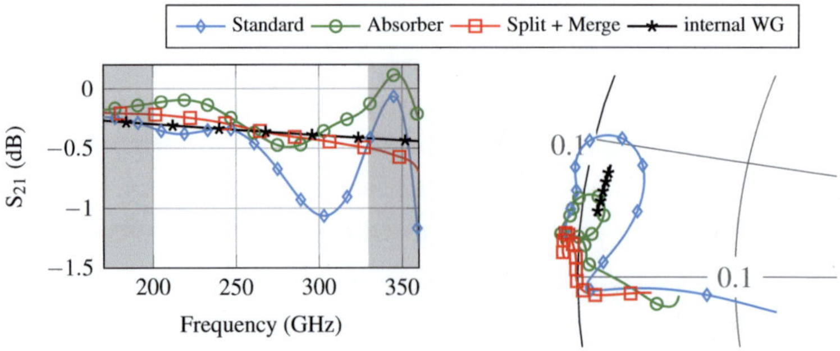

(a) $|S_{21}|$ line (b) S_{11} short

Figure 5.10: EM simulated comparison of different simulation techniques for analyzing the local minima around 300 GHz.

5.2.4 Verification of optimized on-wafer calibration standards

Since the analysis of the previous chapter showed that the remaining deviations cannot be further suppressed, the calibration structure with a closed metallic surface and additional vias promised the best achievable performance and was therefore manufactured and characterized. A chip photograph of the processed MMIC is shown in Fig. 5.11a.

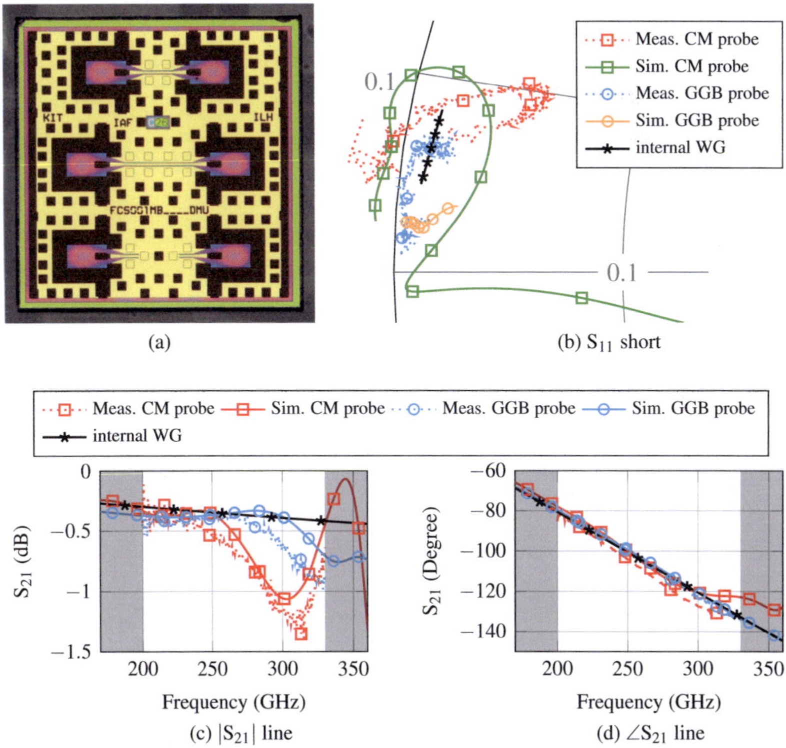

Figure 5.11: (a) Microscopic photograph of the manufactured optimized TRL on-wafer calibration structure. The chip size is $1 \times 1\ \mathrm{mm}^2$. (b), (c) and (d) comparison of the measured and EM simulated optimized on-wafer calibration standards.

In Fig. 5.11b-d the resulting S-parameters of the calibration standards are shown after an on-wafer calibration - with reference planes 150 µm behind the pad - were applied. It can be seen that the measurements closely follow the predicted behaviour based on the EM simulations.
Evaluating the reflection coefficient S_{11} of the short standard shows that using the GGB probe results in a more concentrated behaviour in the Smith diagram, which is closer to the behaviour predicted by the internal WG simulation. The measured transmission coefficient S_{21} shows the predicted minimum around 300 GHz when the standards are measured with the CM probe. In case the standards are contacted with the GGB probe the minimum is less distinct and shifts up in frequency towards 330 GHz. Similar behaviour can be seen in the phase of S_{21} where the CM probe shows small distortions towards 330 GHz but the GGB probe gives a flat decrease of phase over frequency, as predicted by the internal WG simulation. The lower distortions of the GGB probe are a direct result of the lower RF probe crosstalk, as already discussed in section 3.2.

5.3 Conclusions on chapter 5

In this chapter the concept of on-wafer calibration, which promises to diminish the influence of the RF probe and contact pad, was discussed in detail. However, the results indicate that even though an advantage compared to ISS calibrations can be seen, several parasitic effects remain. The analysis of the existing on-wafer calibration standards showed several resonances in the frequency range of interest, which distort the calibration accuracy and therefore the measurement results. By using the RF probe models and EM simulations, these resonances have been traced back to parasitic modes at the edge of the MMIC and coupling into adjacent standards. To suppress these parasitic modes the developed RF probe models were used to design optimized calibration standards. The optimization of the on-wafer calibration standards was carried out in several steps. First the on-chip environment of the standards was investigated showing that a closed metallic surface with a large amount of vias leads to the most robust behaviour. Next, it shows that the majority of the remaining deviations from the ideal behaviour are a result of the direct crosstalk between the RF probes and cannot simply be resolved by extending the spacing or length of the standards, within reasonable limits. Therefore, the calibration substrate with a closed metallic surface and a large amount of vias was chosen for realization and a very good agreement between simulation and measurement was achieved. Even though the optimized standards show the influence of the crosstalk this can very well be predicted and methods for correction will be discussed in chapter 6.

6 Probe influence on MMIC performance

Up to this point, the influence of the probes was investigated by simple contact pad test structures and ISS calibration standards as well as on-wafer calibration standards. However, while these DUTs are suited for verification of the proposed methods they have no sophisticated functionality and are therefore of limited use in practical systems. To investigate the influence on actual system components, a broad variety of different DUTs was designed, fabricated and characterized. For verification of passive circuit topologies, i.e. without amplifying circuit elements, three different circuit topologies were investigated: reflective-type phase shifter (RTPS), single pole double throw (SPDT) switch and digital phase shifter (DPS). Using phase shifters as verification devices has the advantage, that not only the magnitude but also the phase is of interest, which is not considered in most other system components. The passive nature of these concepts furthermore results in several dB of insertion loss, which makes these devices sensitive to the influence of probe crosstalk and therefore ideal for the investigations of such effects. To investigate amplifying MMICs, two different topologies were chosen: active power divider (APD) and balanced variable gain amplifier (VGA). For the APD, several test circuits with different pads using the same core circuit were designed, allowing to analyse the pad dependent influence on the circuit performance. While the APD is a low gain device, the designed VGA, realized with the PDK bond-pad, as well as with the optimized bond-pad, has a high gain above 20 dB and is based on a three-stage cascode amplifier design, completing the range of investigated topologies.

6.1 Crosstalk influence on the transmission coefficient

Before actual DUTs are evaluated it is important to investigate how and at which magnitude parasitic effects will distort the measurement results. As already discussed, RF probes are prone to excitation of higher order modes, crosstalk between probes, coupling to adjacent structures, radiation into the substrate etc. While coupling to adjacent structures and radiation into the substrate can be minimized by a proper design of the pads and calibration standards, direct crosstalk between probes remains a challenging issue. One approach to capture and correct this effect is to account for the crosstalk during the calibration and correct the resulting errors. While the published 12-term and 16-term [But+91] error models are in principle able to correct crosstalk between probes, for many applications this rather results in a degradation than in an improvement of the measurement results accuracy [Wil+14b]. An obvious reason for the degradation is the change of measurement environment i.e. different substrates for calibration standards and DUT when calibrating with an ISS. But even if the calibration standards and the DUT are realized on the same substrate and make use of the same RF pads, more subtle changes such as a changing distance between the probe tips or the use of DC probes for bias supply will affect the crosstalk behaviour. To comprehend these effects, the EM simulation of MMICs including probe tip models allows to investigate the electric fields and S-parameters at reference planes which are not accessible in measurements. This technique allows new insights into the internal behaviour of the circuits and enables the crosstalk correction for arbitrary combinations of DUT and measurement environment [Mül+18a].

6.1.1 Crosstalk model and influence

For evaluating the crosstalk influence on the DUT, the approach of the two-tier interior crosstalk model, which was presented in [Wil+14b] was utilized. The adapted model, showing the corresponding elements of the crosstalk matrix C is shown in Fig. 6.1.

The concept requires that the VNA is first calibrated with a 7-term calibration procedure without crosstalk correction such as the TRL or LRRM method. This so called base calibration is performed by the error terms E_1 and E_2, as shown in Fig. 6.1 and corrects the results to the preselected reference plane. The crosstalk behaviour is modelled by a intermediary four-port matrix, which matrix elements inherit several simplifications resulting due to the base calibration.

Port 1 to port 3 and port 2 to port 4 are connected transparently (i.e. $C_{11} = C_{22} = C_{33} = C_{44} = 0$ and $C_{13} = C_{31} = C_{24} = C_{42} = 1$). Furthermore, due to the base calibration, the crosstalk terms are reciprocal (i.e. $C_{12} = C_{21}$, $C_{34} = C_{43}$, $C_{14} = C_{41}$ and $C_{23} = C_{32}$) [Wil+14b].

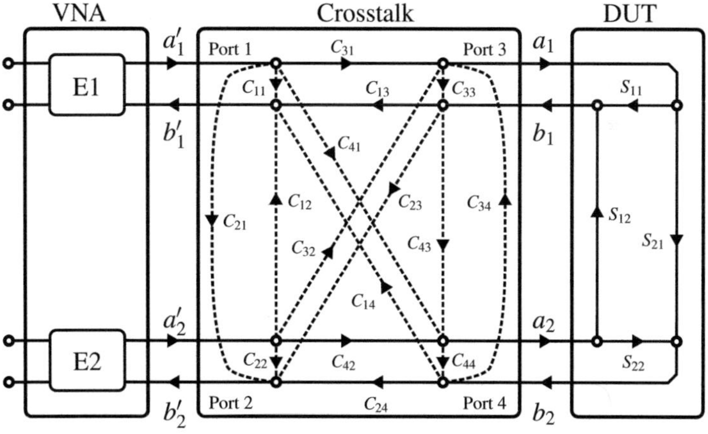

Figure 6.1: Two tier interior crosstalk model used for analyzing the crosstalk [Mül+18a].

Analysing the signal flow graph in Fig. 6.1 and neglecting signal terms with multiple reflections, which is a valid assumption due to usually very low magnitude levels, the measured S-parameter transmission coefficient defined as

$$S'_{21} = \left.\frac{b'_2}{a'_1}\right|_{a'_2=0} \qquad (6.1)$$

results in

$$S'_{21} = S_{21} + C_{21} + C_{23}S_{11} + C_{41}S_{22} + C_{43}S_{11}S_{22}. \qquad (6.2)$$

S'_{21} represents the crosstalk distorted output signal as measured by the VNA and S_{21} represents the crosstalk-free behaviour of the DUT. For a practical circuit, which is fairly matched (6.2) is simplified and the influence on S'_{21} is dominated by the direct crosstalk C_{21} between the RF probes:

$$S'_{21} = S_{21} + C_{21} = |S_{21}|e^{j\varphi_{21}} + |C_{21}|e^{j\phi_{21}}. \qquad (6.3)$$

$|S_{21}|$ and φ_{21} represent the magnitude and phase of S_{21} and $|C_{21}|$ and ϕ_{21} the magnitude and phase of the crosstalk matrix element C_{21}, respectively. The deviation E_{21} to the crosstalk free signal S_{21} is given by

$$E_{21} = \frac{S'_{21}}{S_{21}} = 1 + \frac{|C_{21}|}{|S_{21}|}e^{j(\phi_{21}-\varphi_{21})}. \qquad (6.4)$$

Equation (6.4) shows that the deviation introduced by the crosstalk depends on the magnitude ratio $|C_{21}|/|S_{21}|$ and the phase difference $\phi_{21} - \varphi_{21}$ between S_{21} and C_{21}. The maximum magnitude deviation of (6.4) occurs, if the DUT transmission coefficient S_{21} and the crosstalk coefficient C_{21} are either in phase or 180° out of phase, i.e.

$$\phi_{21} - \varphi_{21} = n\pi \text{ with } n = \{0;1;2;\ldots\} \qquad (6.5)$$

Inserting (6.5) in (6.4) results in the maximum magnitude deviation $|E_{21}|_{max}$ of the transmission coefficient S_{21}:

$$|E_{21}|_{max} = 1 \pm \frac{|C_{21}|}{|S_{21}|} \tag{6.6}$$

Similar considerations can be carried out for the phase deviation introduced by the crosstalk signal. Evaluating the argument of (6.4) gives the corresponding relation between ϕ_{21} and φ_{21} for maximum phase deviation:

$$\phi_{21} - \varphi_{21} = (2n+1)\pi + \cos^{-1}\left(\frac{|C_{21}|}{|S_{21}|}\right) \tag{6.7}$$

Using (6.4) with (6.7) results in the maximum phase deviation:

$$\arg(E_{21})_{max} = \tan^{-1}\left(\frac{|C_{21}|}{\sqrt{|S_{21}|^2 - |C_{21}|^2}}\right) \tag{6.8}$$

Fig. 6.2 evaluates the maximum magnitude and phase error given by (6.6) and (6.8) for different signal and crosstalk magnitudes e.g. a DUT transmission magnitude of -10 dB and a crosstalk magnitude of -30 dB will result in a maximum magnitude and phase error of 0.81 dB and 5° in the measured signal. As can be seen, the crosstalk is primarily affecting lossy devices while showing only minor influence on high gain devices such as amplifiers.

6.1.2 Crosstalk of RF probes

To estimate the probe crosstalk, the magnitude of S_{21} at RF probe tip distances d of 100, 500 and 1000 µm and on different substrates was measured. Additionally, the crosstalk was also verified by EM simulations using the CM probe model presented in chapter 3. The simulation results were corrected by an ISS TRL error set with reference planes at the probe tip. A ruler embedded on the ISS was used to adjust the distance between the probes. To measure the crosstalk in air, the probes were raised several mm above the underlying substrate to neglect the substrate influence. The measured and simulated results are shown in Fig. 6.3a. As expected the crosstalk increases with increasing frequency and decreases with increasing distance. At a distance of 100 µm the crosstalk is below −40 dB for frequencies up to 240 GHz and increases to −25 dB at 330 GHz.

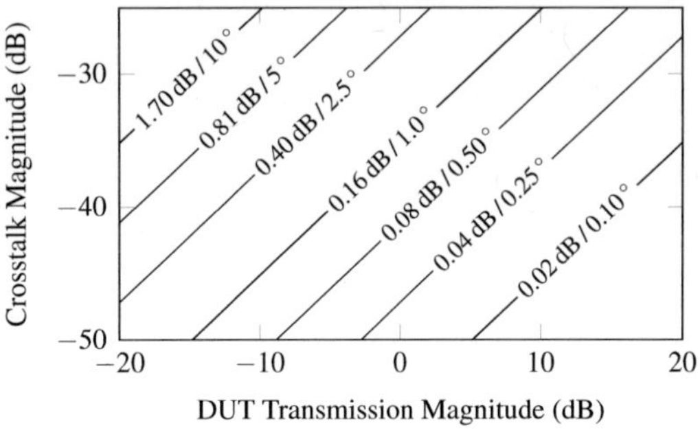

Figure 6.2: Magnitude and phase of crosstalk distortion E_{21} of an arbitrary signal [Mül+18a].

For measuring the crosstalk on alumina, the ISS substrate with a thickness of 254 µm was used. In order to suppress possible substrate modes the ISS was placed on an millimeter-wave absorber, as recommended by the manufacturer. The behaviour of the measured and simulated results is similar to the crosstalk in air and the results are shown in Fig. 6.3b. The crosstalk characteristics on metal differs significantly to previous results, see Fig. 6.3c. At frequencies around 200 GHz a crosstalk of −25 dB is observed while decreasing towards higher frequencies. These results clearly show that, depending on the environment, significant crosstalk between both probes occurs, which can influence the measured behaviour of calibration standards and DUTs.

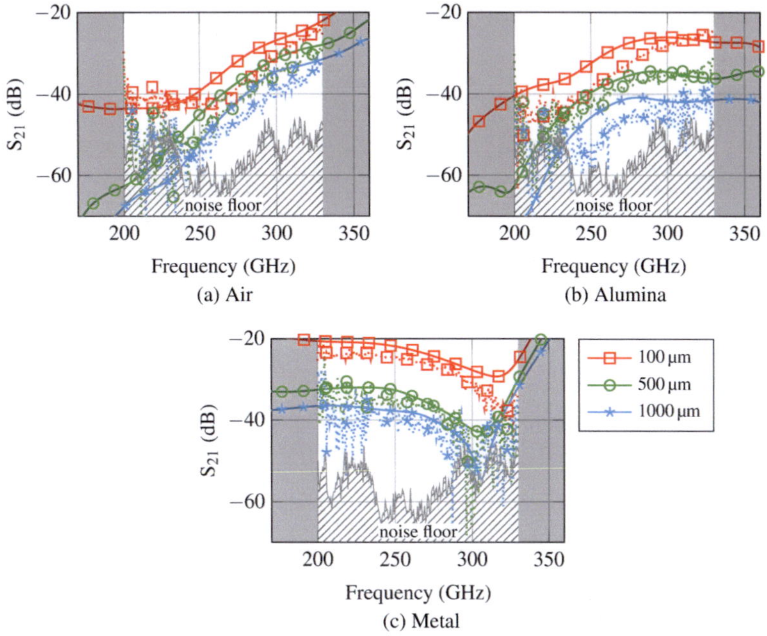

Figure 6.3: Measured (dotted) and simulated (solid) crosstalk using CM probes at different probe distances and on different substrates [Mül+18a].

6.1.3 Crosstalk correction

The magnitude and phase of the crosstalk depend on various conditions of the measurement environment, e.g. the probe distance which in turn is dependent on the chip size, the type of RF probe used for the measurement, DC probes or bond wires required for biasing of the circuit, the support material on which the DUT is placed, etc. Based on error term models used for the 15-term calibration, an enhanced model was developed, which allows the crosstalk correction with respect to the measurement environment. As can be seen in Fig. 6.4 the DUT is split into a core DUT two-port embedded into a four port accounting for crosstalk and probe to pad interconnect effects. To calibrate the measurement system between the VNA and the probe tips, a 7-term calibration is used represented by the two error two-ports E_1 and E_2.

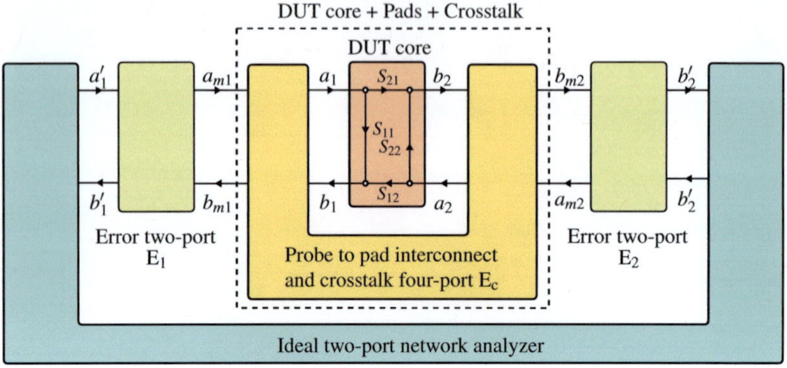

Figure 6.4: Crosstalk correction model for a two-port network analyzer.

The measured (S_m) and crosstalk corrected (S_c) S-parameters are defined as:

$$\begin{bmatrix} b_{m1} \\ b_{m2} \end{bmatrix} = \mathrm{S_m} \begin{bmatrix} a_{m1} \\ a_{m2} \end{bmatrix}, \qquad \mathrm{S}_m = \begin{bmatrix} S_{11m} & S_{12m} \\ S_{21m} & S_{22m} \end{bmatrix} \tag{6.9}$$

$$\begin{bmatrix} b_1 \\ b_2 \end{bmatrix} = \mathrm{S_c} \begin{bmatrix} a_1 \\ a_2 \end{bmatrix}, \qquad \mathrm{S}_c = \begin{bmatrix} S_{11} & S_{12} \\ S_{21} & S_{22} \end{bmatrix} \tag{6.10}$$

The relation between S_m and S_c can be written using scattering transfer T-parameters defined according to Fig. 6.4:

$$\begin{bmatrix} b_{m1} \\ b_{m2} \\ a_{m1} \\ a_{m2} \end{bmatrix} = T \begin{bmatrix} b_1 \\ b_2 \\ a_1 \\ a_2 \end{bmatrix} \tag{6.11}$$

$$T \triangleq \begin{bmatrix} T_1 & T_2 \\ T_3 & T_4 \end{bmatrix} = \left[\begin{array}{cc|cc} t_0 & t_1 & t_4 & t_5 \\ t_2 & t_3 & t_6 & t_7 \\ \hline t_8 & t_9 & t_{12} & t_{13} \\ t_{10} & t_{11} & t_{14} & t_{15} \end{array}\right] \tag{6.12}$$

Inserting (6.9) and (6.10) in (6.11) and solving for S_c results in

$$S_c = (T_2 - S_m T_4) \cdot (S_m T_3 - T_1)^{-1} \tag{6.13}$$

which allows to calculate the crosstalk free behaviour S_c of the DUT. In contrast to the crosstalk analysis of section 6.1.1 neither the reflection coefficients of the four-port E_c can be set to zero nor the transmission elements can be set to unity, which increases the complexity of the analysis. To determine the values of the four-port E_c would require at least five different calibration standards such as thru, open-open, open-match, match-open and match-match [But+91]. Furthermore, to achieve accurate results these standards need to be embedded in the same measurement environment as the DUT, i.e. using the same substrate and identical pads of the DUT with same probe distance and DC bias configuration. For MMIC on-wafer measurements this would require six times the chip area of the DUT itself, which is for almost all applications not practical. However, by setting up an accurate simulation model using EM simulations it is possible to determine the matrix elements of E_c, representing the measurement environment with RF and DC probes as close as possible. Fig. 6.5 shows, by way of example, the simulation setup of a digital phase shifter. The signal at the input and output is excited at the coaxial line of the probe tip model and the second port is placed at the intended reference plane. The simulated S-parameters are calibrated using simulated error terms based on the same calibration procedure and standards which are used in the measurement, i.e. depending on the DUT an ISS or on-wafer calibration. After this calibration the remaining part of the calibrated simulation results is equal to the crosstalk four-port E_c, which can then be applied to the measurement results as a two-tier calibration.

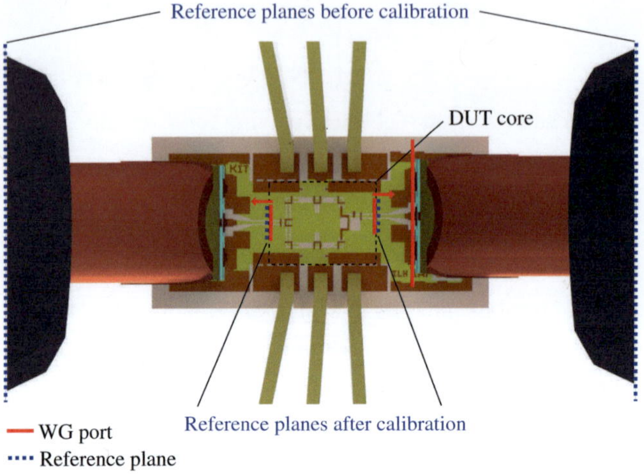

Figure 6.5: Practical example of the DUT core area and reference planes used for simulation of the four-port error matrix E_c.

6.2 Reflective-type phase shifter

The RTPS is an ideal DUT for evaluation of the proposed EM simulation methods, since the magnitude as well as the phase behaviour is of interest. A detailed explanation of the RTPS working principle can be found in appendix C.1.1. Figures of merit (FOM) such as the relative phase shift or RMS amplitude and phase errors are calculated according to C.1. For the design of a RTPS it is inevitable to have access to accurate measurements or models of the used varactor diodes. Therefore, before designing the RTPS, corresponding varactor test structures were realized using the Fraunhofer IAF 50 nm technology [Leu+07].

6.2.1 Varactor measurements

The test structure for the measurement of the varactors was designed using the characterization-pads presented in section 4.1. To shift the reference plane to the varactor diode and precisely extract the device behaviour, an on-wafer calibration was performed using the PDK calibration standards, as well as the optimized on-wafer calibration standards of section 5.2. The measurement results of a 15 µm long varactor for both calibrations is shown in Fig. 6.6 at different varactor voltages V_c. As can be seen, the magnitude and phase of the reflection coefficient S_{11} have a smoother response if the calibration is based on the optimized calibration standards. Close inspection of the magnitude reveals that the resonances of the PDK calibration standards, discussed in section 5.1, are directly distorting the reflection coefficient of the measured varactor data.

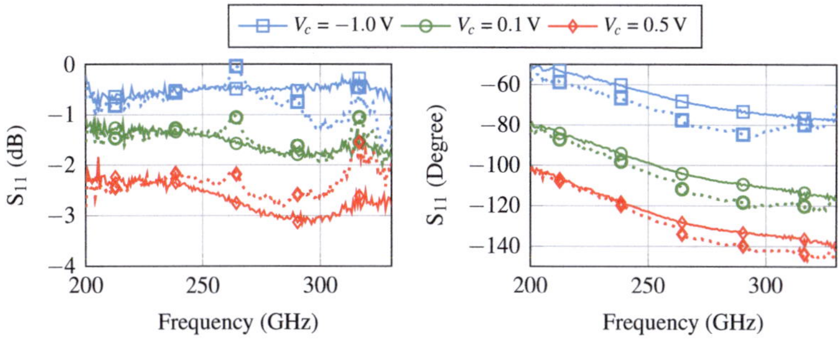

(a) $|S_{11}|$ (b) $\angle S_{11}$

Figure 6.6: On-wafer calibration comparison using PDK (dotted) and optimized (solid) calibration standards for a varactor diode.

6.2.2 RTPS design and measurements

Based on the varactor measurements the design of a RTPS for operation in the WR3 frequency range was carried out. The goal was to achieve a phase shift of at least 120° at a center frequency of 240 GHz, with low loss and broadband operation in terms of RMS phase and amplitude errors. If successfully realized, these specifications would allow to cascade the RTPS into a three stage design addressing the full 360° phase control range. To investigate the influence of the contact pads, the same circuit core was realized with two different pad structures. A large pad with a ground to ground spacing of 80 µm and an inner conductor width of 22 µm was used for one realization of the RTPS. The second design used the PDK measurement-pad presented in 4.1, which should have a much lower influence on the circuit performance. A microscopic picture of both RTPS MMICs is shown in Fig. 6.7. Due to the small dimensions and the requirement of DC probes for bias supply the DUT can only be measured with CM probes. The measurement setup was calibrated using a TRL calibration based on ISS standards and the reference planes were shifted to the probe tips.

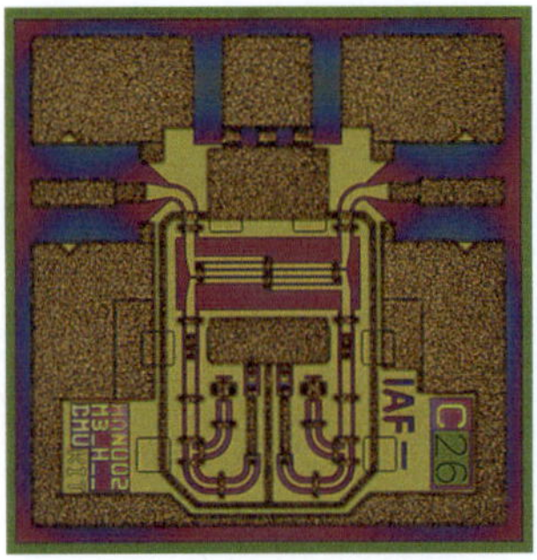

(a)

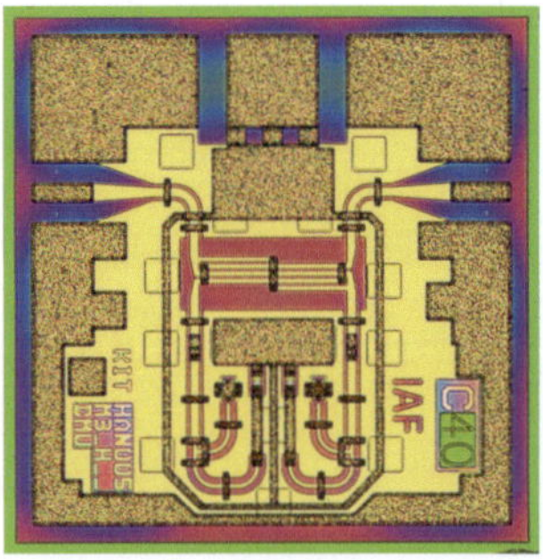

(b)

Figure 6.7: Microscopic picture of the RTPS with (a) large pads and (b) PDK measurement pads. The chip size is $0.5 \times 0.5\,\mathrm{mm}^2$.

Influence of varactor data

Important for the design of a RTPS is the accuracy of the used varactor models. The evaluation of section 6.2.1 showed a dependency of the varactor data set to the used calibration substrate. To investigate the influence of the different varactor data sets, the measurement results of the RTPS design utilizing the measurement-pads were compared to EM simulations of the MMIC using WG excitation as well as simulations including the RF and DC probe tips.

The results are summarized in Fig. 6.8 for typical phase shifter characteristics such as insertion loss, relative phase shift as well as RMS amplitude and phase errors. The control voltage V_c of the RTPS was swept from -1 V to 0.5 V with a step width of 0.1 V. Considering the averaged input reflection coefficient S_{11} a very good agreement is observed between measurement and corresponding RF probe tip simulation. However, there is a large deviation from the matching predicted by the WG simulation. Due to the circuit topology, the use of differently calibrated varactor data shows only minor influence on the input reflection coefficient S_{11}.

The transmission performance on the other hand shows a direct dependency on the used calibration. Evaluating the transmission coefficient S_{21} in Figs. 6.8b and 6.8c shows several deviations between measurement and simulation. While the narrow band distortions can be traced back to the resonances in the PDK calibration standards, the general difference between measurement and simulation cannot be explained by these issues. The decrease in magnitude towards 300 GHz can be attributed to the use of the measurement-pad as discussed in section 4.1 and is therefore no characteristic of the RTPS core but of the contact pad. Evaluating the relative phase shift of S_{21} clearly shows the distortions resulting from the PDK calibration standards. The use of the varactor data acquired with the PDK calibration standards predicts a too low relative phase shift with several distortions above 260 GHz. The varactor data based on the optimized calibration standards predicts a smoother behaviour over the full frequency range and achieves a better match to the measurement results. The relative phase of the EM simulations based on WG excitation and RF probe tip models are in very close agreement, proving that the varactor diodes are successfully isolated from the influence of the RF probe to pad interconnect.

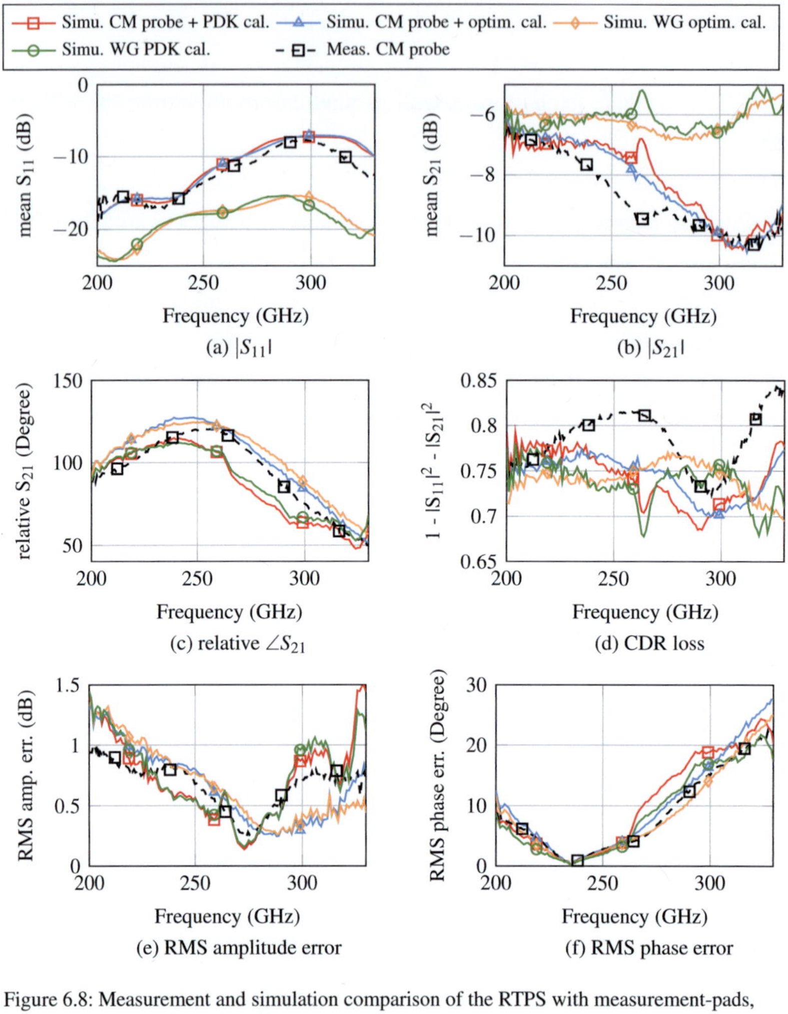

(a) $|S_{11}|$

(b) $|S_{21}|$

(c) relative $\angle S_{21}$

(d) CDR loss

(e) RMS amplitude error

(f) RMS phase error

Figure 6.8: Measurement and simulation comparison of the RTPS with measurement-pads, using differently calibrated varactor data.

The deviations of S_{21} also affect the overall CDR loss of the RTPS and the measured loss is higher than predicted by simulations, as shown in Fig. 6.8d. Evaluating the broadband characteristics such as the RMS amplitude error in Fig. 6.8e and the RMS phase error in Fig. 6.8f generally shows a good agreement of the simulations to the measurements although several small deviations remain. The RMS amplitude error deviates from the simulation using the varactor data based on the optimized calibration standards and shows a better fit to the data obtained with the PDK calibration standards. The RMS phase error on the other hand is very precisely predicted by the varactor data obtained by the optimized calibration standards and deviates for the PDK calibration above 260 GHz. Altogether, the simulations based on varactor data using the optimized calibration standards provide a better match to the measured results.

Influence of contact pad design

For the evaluation of the contact pad influence on the RTPS circuit performance only the varactor data obtained by the optimized calibration standards was used. Again, the two MMICs were fully EM simulated using WG ports as well as the RF and DC probe tips. The measurement and simulation results are summarized in Fig. 6.9. While using the large pads results in an improved input reflection coefficient S_{11}, this does not improve the transmission coefficient S_{21}, as shown in Figs. 6.9a and 6.9b. Evaluating S_{21} in more detail shows that the use of the large pad results in several resonances in the WR3 frequency range, deteriorating the magnitude and even the phase behaviour (Fig. 6.9c) of the RTPS. The overall loss, which is compared in Fig. 6.9d, is also strongly affected by the different pads. Further deterioration can be observed by evaluating the RMS amplitude and phase errors in Figs. 6.9e and 6.9f, respectively. However, while the EM simulations using WG excitation fail to predict these effects they can be predicted precisely by including the RF probe tip models into the simulation. This allows to predict precisely the actual circuit behaviour and cuts down the number of necessary design iterations.

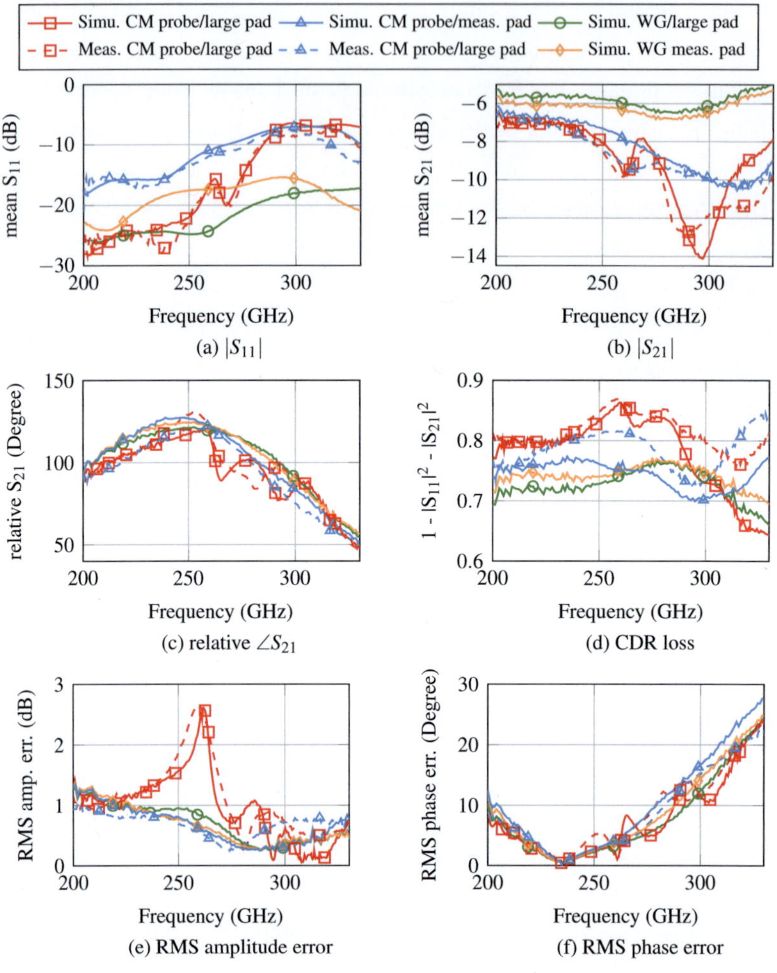

Figure 6.9: Comparison of measured and simulated data of two RTPS using different pads.

6.3 Digital phase shifter

The verification of the RF probe crosstalk influence was performed with a 180° DPS MMIC processed in the Fraunhofer IAF 50 nm technology. Details on the phase shifter design can be found in appendix C.1.2. A microscopic photograph of the processed MMIC is depicted in Fig. 6.10, additionally showing the on-wafer reference planes.

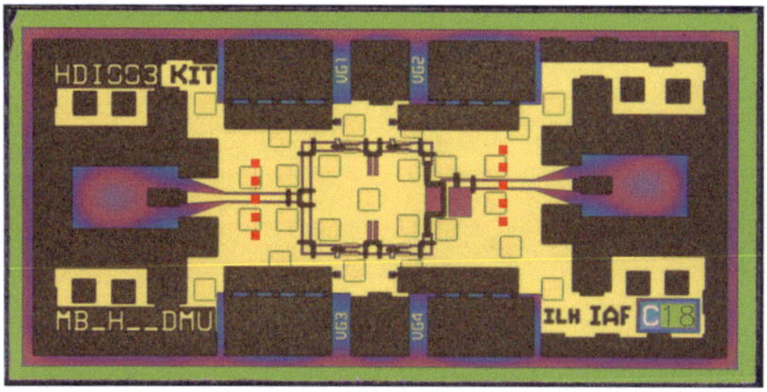

Figure 6.10: Chip photograph of the 180° DPS MMIC with size of $1.0 \times 0.5\,\text{mm}^2$. The reference planes are shown as red dashed lines.

The total chip size is $1.0 \times 0.5\,\text{mm}^2$ and the RF probe tips are placed approximately 550 µm apart from each other. Due to the phase shift of 180° in the phase of the transmission coefficient S_{21}, the errors predicted by (6.6) and (6.8) double when evaluating the magnitude and phase imbalance between both switching states. This makes 180° phase shifters extremely sensitive in terms of crosstalk, but also an ideal candidate for investigating such effects. To analyse the crosstalk influence on the used phase shifter, four different scenarios shown in Fig. 6.11 were investigated. Fig. 6.11a shows a 3D EM model of the DPS MMIC contacted with CM probes on the input and output pad. For this scenario no DC bias supply is modelled, which is not possible in an actual measurement. In Fig. 6.11b this scenario is extended by a model of the DC probes, which is one possibility for bias supply in measurements. Alternatively, bondwires can be used for bias supply, as shown in Fig. 6.11c. To verify the crosstalk free behaviour, an additional scenario was simulated with signal excitation by WG ports directly at the reference plane of the on-wafer calibration, see Fig. 6.11d. Furthermore, the reference planes for connecting the transistor models in the post-processing step, are additionally shown in Fig. 6.11d, but are also used in the other scenarios. All reference planes are shown as red dotted lines. Simulations verified that in case WG ports are used for signal excitation, the type of bias supply has no influence on the simulation results, which indicates that the crosstalk originates solely from the RF probes.

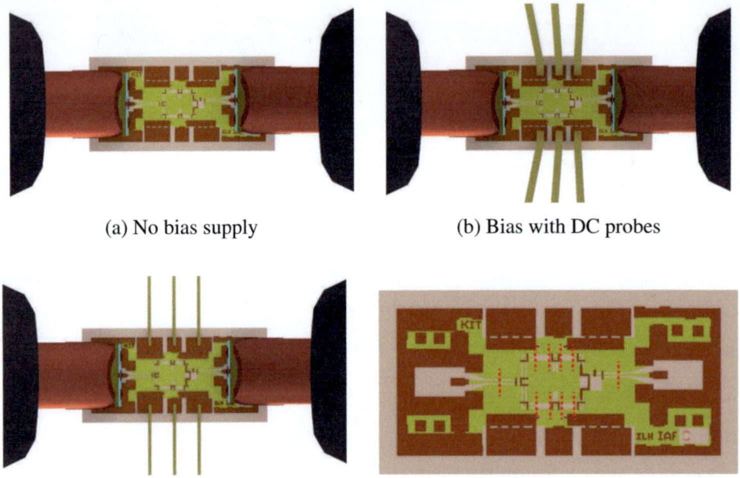

(a) No bias supply

(b) Bias with DC probes

(c) Bias with bondwires

(d) Close-up MMIC EM simulation model

Figure 6.11: Overview of the different simulation scenarios for the discrete 180° phase shifter.

For verifying the simulations, the scenarios with DC probe and bondwire bias supply were realized and measured. The measurement and simulation results of the different scenarios are shown in Fig. 6.12. Compared to the simulations including the RF probe tips, the measured results show almost identical behaviour versus frequency but at a higher mean insertion loss in S_{21} of approximately 1 dB, see Fig. 6.12a. In contrast to the measurements, as well as the RF probe tip simulations, the WG port simulation shows a flat

S_{21} response over the full band. A significant difference between measurements and RF probe tip simulations to the WG port simulation can be seen in the magnitude and phase imbalance between both switching states, shown in Figs. 6.12b and 6.12c, respectively. While the measurements and simulations using DC probes as bias supply follow the WG simulation closely, the results based on the bondwire DC supply show up to 0.8 dB and 5° imbalance for the magnitude and phase, respectively. The imbalance is even higher for the simulation results without any physical DC bias supply, exceeding 1 dB magnitude and 7° of phase imbalance. These results clearly show the influence of different scenarios on the occurring crosstalk.

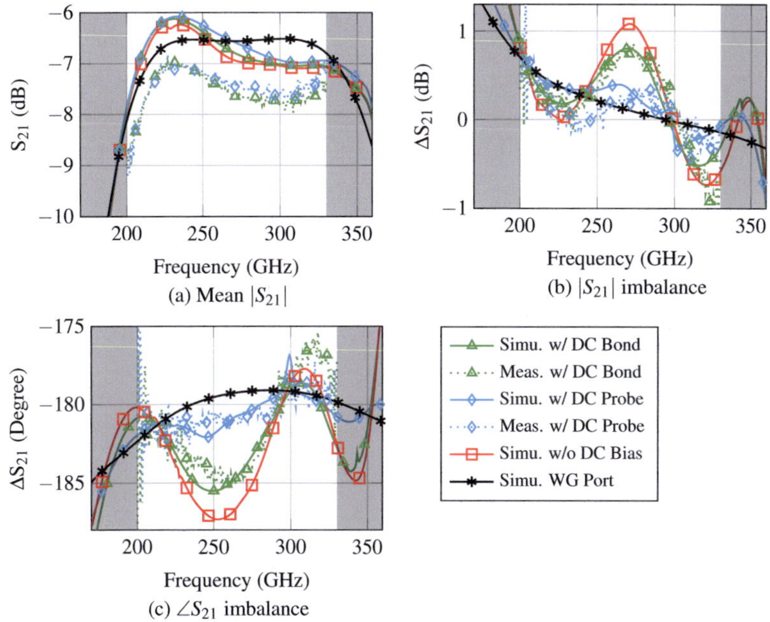

(a) Mean $|S_{21}|$

(b) $|S_{21}|$ imbalance

(c) $\angle S_{21}$ imbalance

Figure 6.12: Measured and simulated S_{21} characteristics of the 180° DPS.

6.3.1 Crosstalk correction for the digital phase shifter

Using the technique proposed in section 6.1.3, it is possible to correct the crosstalk which is distorting the accuracy of the measurement results. To calculate the values of the crosstalk error four-port E_c, WG ports are placed at the reference planes and the direct signal path from RF probe to the reference plane is simulated. After calibrating these results with the corresponding simulated on-wafer calibration error terms, the results represent the crosstalk matrix E_c introduced in section 6.1.3. Evaluating the resulting coupling coefficient C_{21} allows to gain insight in the direct crosstalk between both RF probes, which is the major crosstalk influence. The results are shown in Fig. 6.13 in magnitude and phase. The crosstalk of the different scenarios is clearly separated with magnitudes between a maximum of −28 dB without physical bias supply and a minimum of −51 dB with DC probe bias supply.

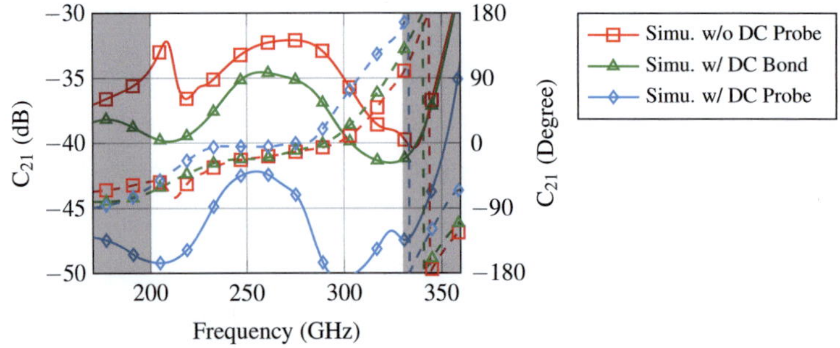

Figure 6.13: Simulated crosstalk C_{21} magnitude (solid) and phase (dashed) of different DC bias configurations.

These results indicate that the DC probes suppress direct crosstalk between the RF probes, which is one of the reasons why calibration algorithms including crosstalk corrections fail when measuring actual DUTs. Next, the simulated crosstalk matrix E_c was used to correct the measurement results of the corresponding scenario. This is performed in Fig. 6.14 and shows that the magnitude and phase imbalance of the different scenarios are corrected to almost identical results, which proves that the different scenarios are precisely captured by the EM simulations. Furthermore, the corrected results are in much better agreement to the ideal WG port simulation, which represents the crosstalk free behaviour of the DUT. As can be seen, the decrease in the magnitude of S_{21} towards higher frequencies was not a characteristic of the circuit but caused by the RF probe crosstalk. Furthermore, the phase deviation from 180° was distorted from the corrected behaviour up to 7°, depending on the DC bias environment.

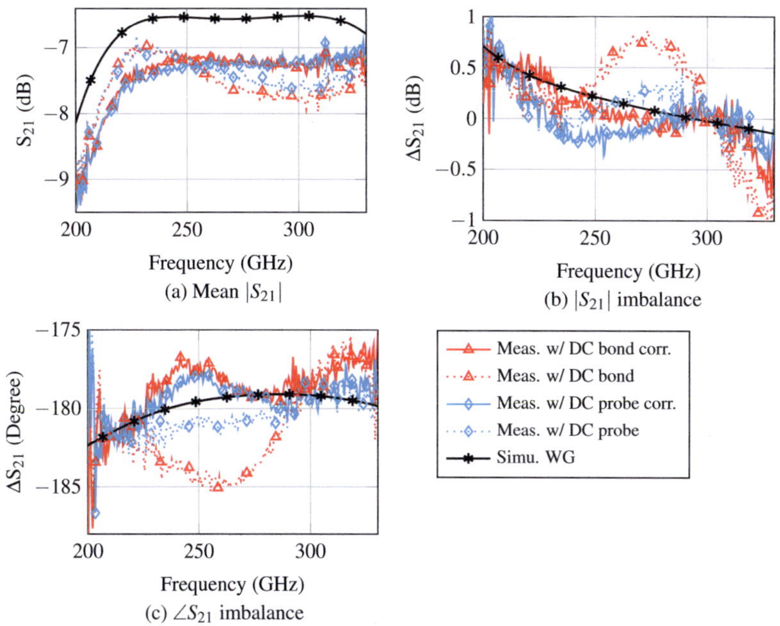

(a) Mean $|S_{21}|$

(b) $|S_{21}|$ imbalance

(c) $\angle S_{21}$ imbalance

Figure 6.14: Corrected and uncorrected measurements and ideal simulation results of a 180° discrete phase shifter.

6.4 Single pole double throw switch

The SPDT switch used by the digital phase shifter of section 6.3 was also realized as a break-out circuit. The detailed description of the operating principle and design of the SPDT switch is provided in appendix C.2. A photograph of the realized MMIC is shown in Fig. 6.15. The utilized contact pads are the measurement-pads presented in section 4.1.2. For the measurements the CM probes were used and the calibration was performed by a TRL calibration using the ISS calibration standards. The reference plane of the calibration was shifted to the RF probe tips.

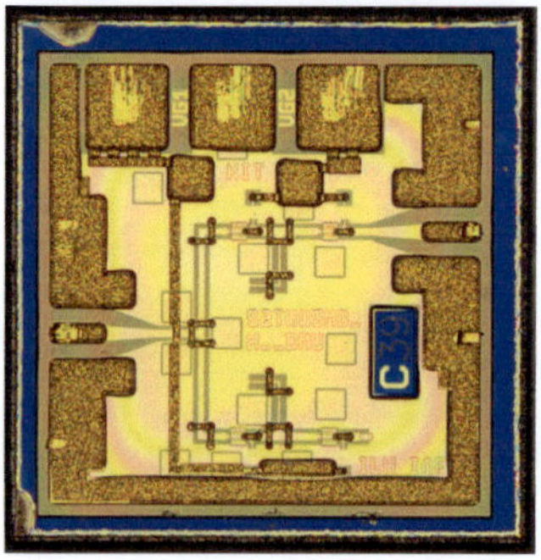

Figure 6.15: Microscopic photograph of the SPDT switch. The chip size is $0.5 \times 0.5\,\mathrm{mm}^2$.

The measurement and simulation results are summarized in Fig. 6.16. The left column shows the results if the upper branch of the switch is biased in the active state, i.e. the signal is passing through. The results for the isolating state of the upper branch are shown in the right column. As can be seen, the input reflection coefficient S_{11} is, to a first order, constant for both switching states. The small differences between both switching states result from the on-wafer $50\,\Omega$ termination, which behaves differently from the termination of the RF probe at the output contact pad. The output reflection coefficient S_{22} behaves as expected with a good matching in the active state and a high reflection for the isolating state.

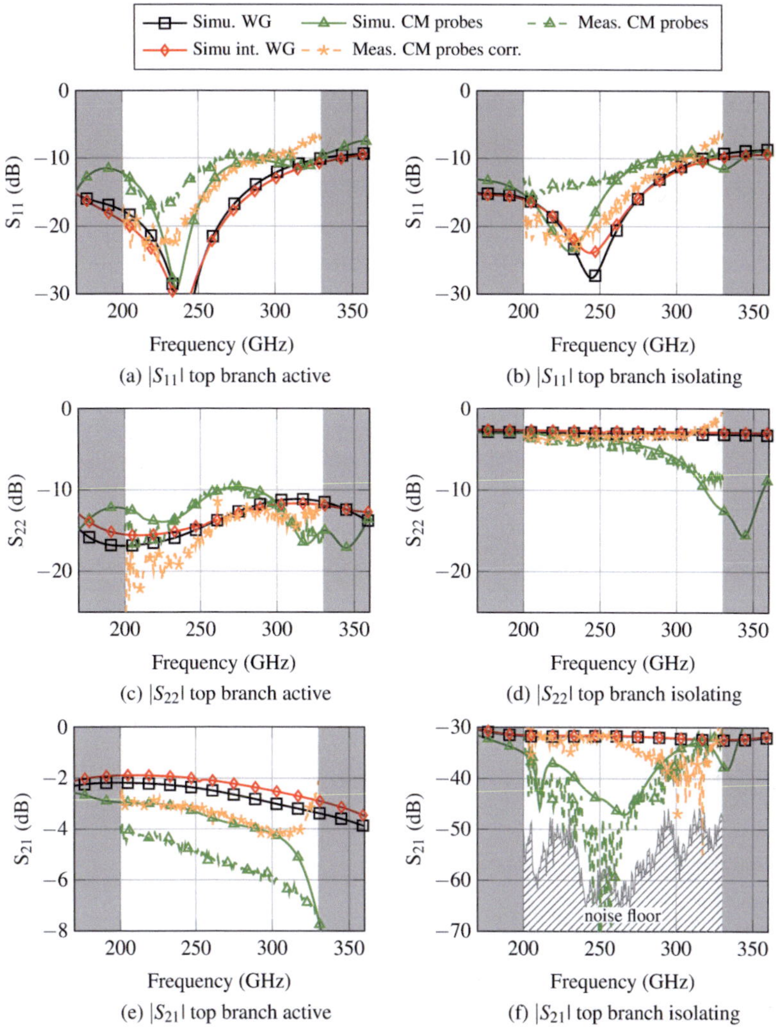

Figure 6.16: Simulation, measurement and corrected measurement results of a SPDT switch.

Evaluating the transmission performance shows a maximum measured insertion loss of 7 dB and a maximum isolation of over 50 dB. Comparing the measured isolation to the simulations using WG ports and RF probe models indicate that the extremely high isolation of 50 dB is only present if probe tips are used. This indicates that the high isolation is a parasitic effect caused by the crosstalk of the RF probes.
To correct this, the same technique presented in section 6.1.3 and already successfully used in section 6.3 was used to apply a crosstalk correction to the measurement. The de-embedding data was created by simulating the MMIC contacted with RF and DC probes. In contrast to the full simulation of the MMIC there were no WG ports at the interfaces to the transistors. WG ports were added at the input as well as at the output of the MMIC at the narrow CPW line, located behind the taper of the contact pad. As can be seen in Fig. 6.16e the shift of reference plane from the RF probe tip to the narrow CPW line decreases the insertion loss of the active state by approximately 1 dB. Furthermore, the corrected measured isolation of the SPDT switch is now in much better agreement with the WG simulated results, proving that the high isolation was not an attribute of the MMIC, but a parasitic effect of the measurement setup and that the real isolation of the switch is around 30 dB. Furthermore, after crosstalk correction the input and output matching are also in much better agreement to the simulated values using WG ports.

6.5 Active power divider

To further verify the EM RF probe simulations and investigate the impact of different contact pads on an active, amplifying circuit an APD topology was chosen for verification [Mül+17b]. A detailed description of the APD topology can be found in appendix C.3. The used pads were the PDK characterization-pad, the PDK bond-pad and the TFMS-pad presented in chapter 4. Since the core circuit uses a CPW environment the TFMS-pad

uses an additional TFMS to CPW transition. Microscopic photographs of the realized MMICs are shown in Fig. 6.17.
For the measurements the CM probes along with a TRL calibration on the ISS with reference planes at the RF probe tip were used. This allows direct evaluation of the influence of the different contact pads to the overall performance. Furthermore, using the PDK characterization-pad allows performing of an on-wafer calibration to remove the influence of the RF probe to pad interconnect and therefore measure the core performance of the APD. The corresponding EM simulation was set up by placing WG ports at the reference planes of the on-wafer calibration.

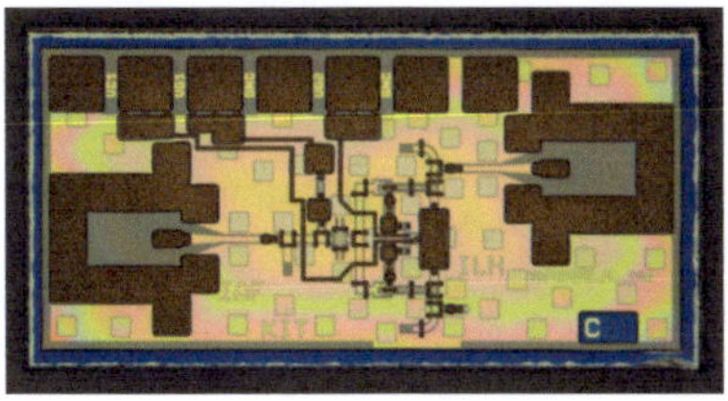

(a) PDK characterization-pad

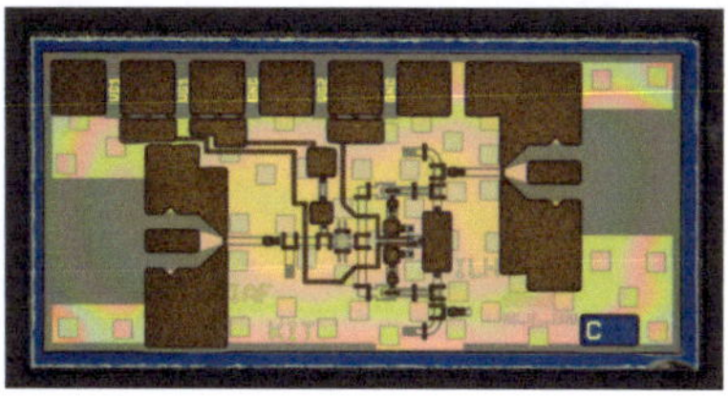

(b) PDK bond-pad

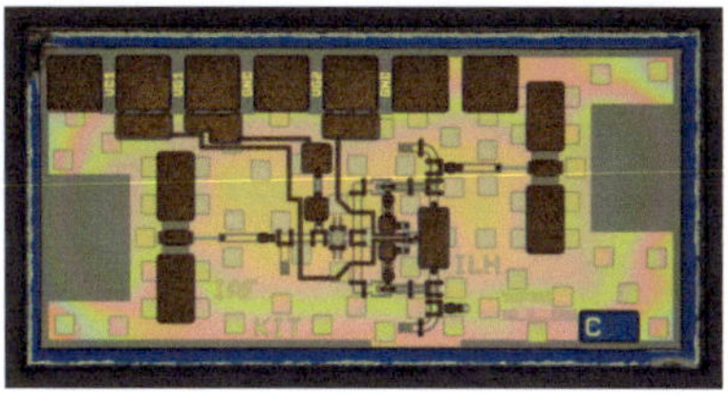

(c) TFMS-pad

Figure 6.17: Microscopic photograph of the active power dividers with different pads. The chip size is $1.0 \times 0.5\,\mathrm{mm}^2$.

The measurement and simulation results are summarized in Fig. 6.18. Evaluating the input and output reflection coefficient S_{11} and S_{22} shows that the influence of the different contact pads is clearly visible, but an advantage of a specific pad cannot be identified, as shown in Figs. 6.18a and 6.18b. However, the use of the PDK bond-pad leads to two resonances in S_{22} around 240 GHz and 295 GHz, which are also predicted by the results of the EM RF probe tip simulation.

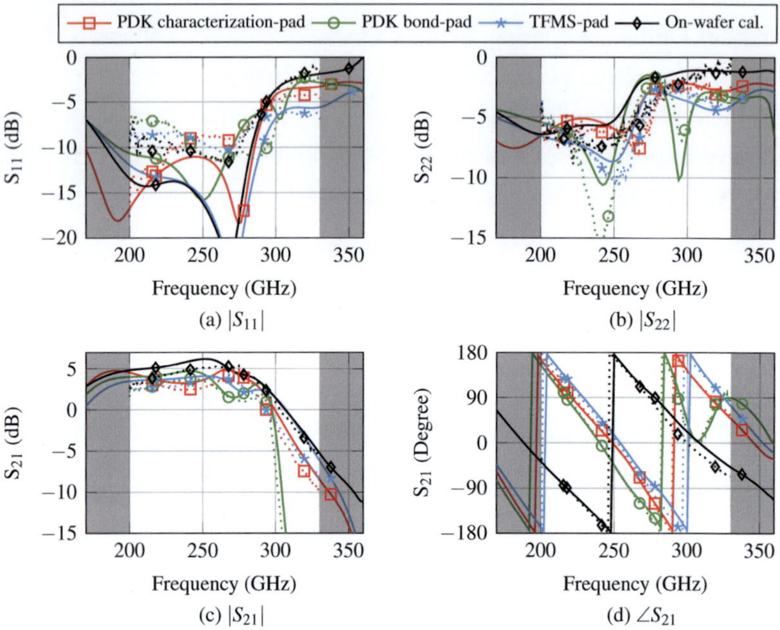

(a) $|S_{11}|$ (b) $|S_{22}|$ (c) $|S_{21}|$ (d) $\angle S_{21}$

Figure 6.18: Simulation (solid) and measurement (dotted) of the APD with different pads.

On the other hand comparing the transmission coefficient S_{21}, shown in Fig. 6.18c, exposes larger deviations between the different pads. Especially the PDK bond-pad results in a sharp decrease in magnitude above 290 GHz, which is in good agreement to the investigation of section 4.1 and is a direct results of the parasitic modes excited in the volume under the pad. The characterization-pad and the TFMS-pad are in close agreement with a slightly higher 3-dB bandwidth of the MMIC using the TFMS-pad. As expected, the on-wafer calibrated results show the lowest loss and highest

3-dB bandwidth. The distortions introduced by the PDK bond-pad are also visible in the phase of S_{21} where the resonance is clearly visible around 300 GHz. Despite the varying behaviour of the different contact pads the simulated behaviour including the RF probe tips accurately predicts the circuit performance. The remaining deviations, especially in S_{11}, are mostly attributed to the limited accuracy of the used transistor models.

6.6 Balanced variable gain amplifier

To verify the EM probe tip simulations with a high gain device, a three stage balanced VGA was designed, realized and characterized. Since the MMIC was intended for packaging purposes, specific focus was on investigating the difference between the PDK bond-pad and the optimized bond-pad. Therefore, two MMICs were realized using the same circuit core but different pads. The detailed description of the circuit core design can be found in appendix C.4

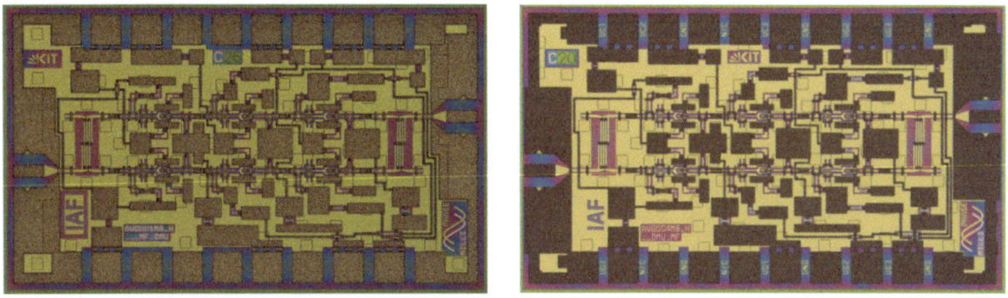

(a) PDK bond-pad (b) Optimized bond-pad

Figure 6.19: Microscopic photograph of the VGA using different pads. The chip size is $1.25 \times 0.75\,\text{mm}^2$.

Due to the balanced amplifier topology a good input and output matching is expected. Microscopic pictures of the two MMICs are depicted in Fig. 6.19, showing that there is only a minor difference between the MMICs regarding the placement of logos and text. The changes made by the optimized bond-pad are not visible in the picture since only the number and position of the vias are changed, as discussed in section 4.3. An ISS TRL calibrated measurement setup was used with CM probes and reference planes shifted to the RF probe tips. For verification, the MMIC was EM simulated with WG port excitation, as well as simulations including the RF and DC probe tips. The measurement and simulation results are presented in Fig. 6.20. The input and output reflection coefficients are summarized in Figs. 6.20a and 6.20b and show, that the use of the PDK bond-pad results in a strong deviation from the WG port simulated and generally expected behaviour of a balanced amplifier. While the measured transmission coefficient is below −10 dB at frequencies around 200 GHz it increases up to −5 dB at 300 GHz. The use of the optimized bond-pads on the other hand results in a much more robust electrical behaviour with a reflection below −10 dB up to 300 GHz and only slight deterioration in the output reflection coefficient towards higher frequencies. The measured gain of the amplifier, shown in Fig. 6.20c, is also affected by the choice of contact pad. Using the PDK bond-pad results in a decrease in magnitude and an earlier gain drop towards higher frequencies. Evaluating the results shows, that the WG simulations give a good estimation of the transmission coefficient S_{21} but do not accurately predict the behaviour of the reflection coefficient S_{11}. In terms of S_{21} phase, a resonance is visible around 300 GHz for the MMIC using the PDK bond-pads, which is in agreement with the earlier discussion about the PDK bond-pad. Generally, it is notable that the phase deviates from the simulations by up to 100°, which is, considering the good agreement between measurement and simulation of earlier results, a large deviation. It is assumed that the phase deviation results from inaccurate reference planes of the used transistor models, which become apparent due to the three-stage design. However, despite the remaining phase difference, using the optimized bond-pads results in an electrically robust behaviour. The issue of worsening input matching towards higher frequencies, which is an unexpected behaviour for a balanced amplifier, is successfully resolved with the optimized bond-pad.

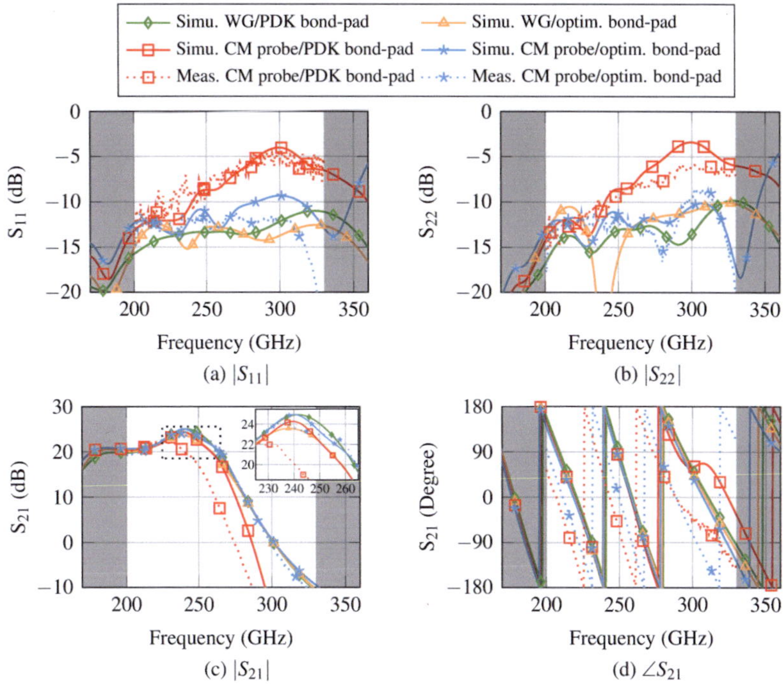

Figure 6.20: Comparison of the performance of the balanced VGA with different bond-pads.

6.7 Conclusions on chapter 6

In this chapter the applicability of EM simulations to complete MMICs including RF and DC probe tips was investigated. The concept was proven successfully for passive circuits such as an RTPS, a DPS and an SPDT switch, as well as active, amplifying circuits such as an APD and a balanced VGA. It has been shown that various effects such as parasitic modes in contact pads have a strong influence on the measured circuit behaviour. Furthermore, crosstalk between the RF probes distorts the measurement results. This effect occurs especially for passive devices due to their inherent loss and therefore low S_{21} magnitude, even if an on-wafer calibration is utilized.

Investigations showed that the crosstalk influence cannot be corrected in the calibration procedure since it is dependent on the measurement environment, i.e. there is a dependency on the probe distance and the type of bias supply used in the measurement setup. However, keeping the parasitic effects low by choice of electrically robust contact pads, the crosstalk can be corrected by advanced EM simulations and de-embedding techniques. It is therefore possible to unveil the true circuit behaviour hidden in a measurement superimposed with parasitic effects. Verification of this approach was successfully carried out for a SPDT switch and a DPS.

7 Conclusion and Outlook

To fulfil the demand on high data rates up to several hundred Gigabit/s, communication systems are pushing into the high mmW and sub-mmW frequency range, where high bandwidths are available. At these frequencies it often happens that measurements do not confirm the expected results, predicted by simulations. Many times this is attributed to fabrication tolerances or inaccurate modelling of the used transistors. However, from the findings of this work, the conclusion can be drawn that many of the observed deviations are accountable to the influence of the RF probes and the measurement setup on the DUT performance.

To eliminate the influence of possible fabrication tolerances or any other unknown influences, the investigations in this work were carried out by means of EM simulations, which give absolute control over the simulation boundaries and are perfectly reproducible. RF probe tip models for the EM simulation are inevitable for carrying out these investigations. The development of the EM simulation probe tip models, deduced from two different commercially available RF probes, was discussed in detail. Furthermore, precise EM simulation models of the DUT MMICs are required, which is challenging due to the non-planarized structure of the Fraunhofer IAF technology. A technique allowing to import such complex non-planarized MMIC layer stacks to commercially available EM simulation software was developed and is presented in this work.

To prove the RF probe tip influence on the DUT performance, this work carried out a comprehensive study by evaluating various DUTs of different topologies. In order to reproduce the measurement as closely as possible and to achieve the same reference planes, the EM simulation results were calibrated using the same calibration procedure used in the measurements. The calibration error terms for the simulation results are based on simulated calibration standards, which are deduced from the real calibration standards. To the best of the authors knowledge this is the first time that such an approach is utilized at mmW frequencies. The deduced probe tip models were first verified using different calibration standards, achieving a very good overall prediction of the measured behaviour. After ensuring the validity of the developed models, the existing contact pad structures were evaluated and several distortions in the frequency range from 200 to 330 GHz were found. These new insights, obtained by the EM simulations of this work, were used to design several new and optimized contact pad structures. A distortion free and low loss operation up to frequencies of 330 GHz was achieved by these pads, which can as of now be used for future MMIC designs. Furthermore, using the EM probe tip simulations, a new set of electrically robust on-wafer calibration standards was designed, which improved the accuracy for the parameter extraction of transistor test structures. In addition to the transmission line test structures and calibration standards, several MMICs containing passive and active circuits were used for verification. For these DUTs, a very good agreement between measurement and simulation was also achieved, unveiling effects which could not be comprehended before this study. The high quality of the developed RF probe models even precisely reproduces low magnitude levels, such as crosstalk between RF probes, which allows to investigate measures for correcting the resulting distortions. It has been shown that even with robust contact pads the occurring crosstalk effects cannot be prevented in on-wafer measurements at mmW frequencies. However, using an adapted two-tier

calibration approach it was possible to successfully correct crosstalk distorted measurements based on simulated correction matrices and the true circuit performance was revealed.
In summary, the techniques and results acquired in this dissertation allow to precisely predict the measured behaviour of mmW circuits by considering the RF probe influence for the first time. This is an important increase in the accuracy of simulating MMICs and builds the foundation for the next level of MMIC design and optimization, without costly and time consuming trial and error iterations.
Especially the design of sub-mmW and terahertz circuits will, due to increasing parasitic effects, require such a considerate verification using probe models in conjunction with EM simulations before being manufactured. While this study was restricted to GaAs MMICs, the investigated effects are not restricted to this technology but will also affect silicon-based MMICs and any other technology operating in the mmW frequency range and beyond. Furthermore, the presented investigations cannot only be applied to the design of MMICs but are also a useful tool for designing high frequency printed circuit boards, which have gained enormous interest in recent years.

A Enhanced probe tip model verification

To further investigate the characteristics of the used RF probes and to verify the deduced EM simulation probe models, additional simulations and measurements of two different scenarios were carried out. The first scenario considered calibrating the probe tips on the ISS of a different manufacturer i.e. to calibrate the CM probes on a GGB calibration substrate and vice versa. In the second scenario the probes are first calibrated using their corresponding ISS and then used to measure the standards on a different ISS.

A.1 Probe calibration on different ISS

The verification of the probe models in section 3.2 showed a good agreement between measurement and simulation of various standards. However, strictly speaking the results are a combination of the probe tip model and the calibration substrate model. During development, the probe tip models were tuned to achieve satisfying results for the corresponding calibration structures. In this process the models of the calibration standards, were assumed to be perfectly remodelled by the EM simulation models, due to their simple layer stack and geometry. However, using this approach it is possible that potential errors in the probe tip and calibration standard models cancel each other out and the results appear to be correct but on a false foundation. To uncouple the development of the probe models and the corresponding ISS, the RF probe models were verified with the ISS of a different manufacturer, which was developed separately.

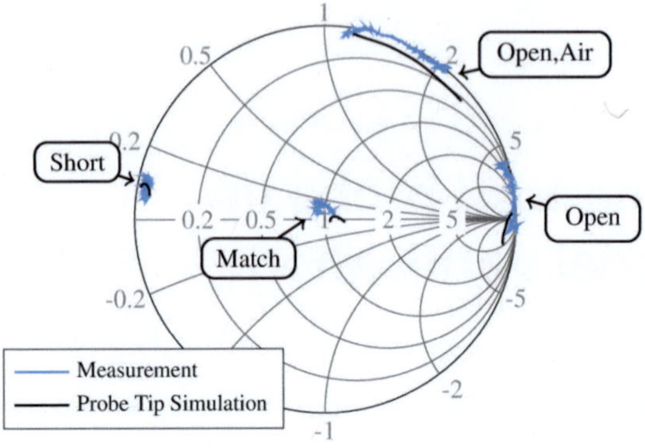

Figure A.1: S_{11} measurement and simulation of Cascade Microtech ISS138-357 calibration standards between 200 to 330 GHz contacted with GGB Picoprobes.

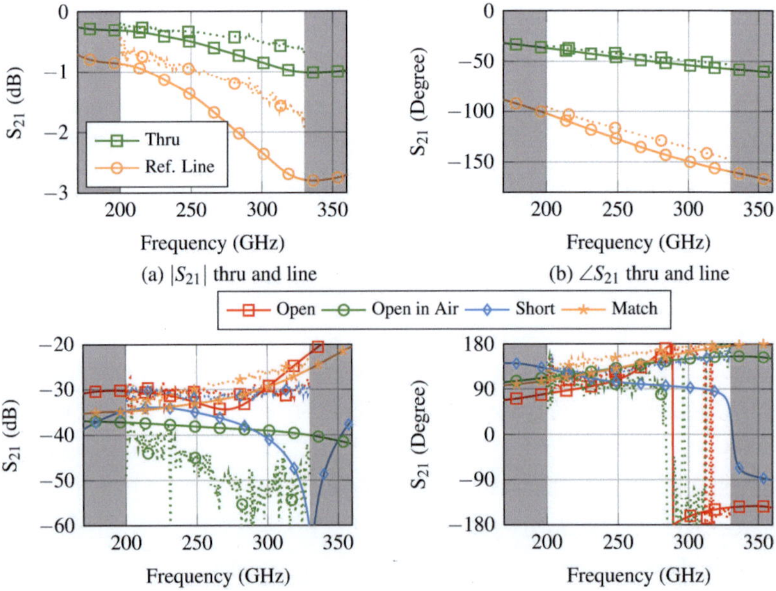

(a) $|S_{21}|$ thru and line

(b) $\angle S_{21}$ thru and line

(c) $|S_{21}|$ reflect and match

(d) $\angle S_{21}$ reflect and match

Figure A.2: Measurement (dotted) and simulation (solid) of S_{21} for different Cascade Microtech ISS138-357 standards contacted with GGB Picoprobes.

Furthermore, the calibration on a different ISS allows to investigate if the resulting characteristics of the calibration standards, such as inductive or capacitive components, is dependent on the utilized probe or calibration substrate. The resulting behaviour of the GGB probe calibrated on the CM ISS138-357 is presented in Figs. A.1 and A.2 for the reflection coefficient S_{11} and transmission coefficient S_{21}, respectively. Comparing the behaviour of the reflection coefficient S_{11} in Fig. A.1 to the results presented in Fig. 3.9 shows, that the location of the standards is comparable to the results of the ISS138-357 when contacted with a CM probe, which suggests that the resulting characteristics are mainly dependent on the ISS and not on the used probe. The transmission and crosstalk behaviour of the thru and line as well as the reflect standards are shown in Fig. A.2. In terms of transmission S_{21} of the thru and line standard an offset of up to 1 dB between measurement and simulation is clearly visible. The measured crosstalk of the reflect standards follows the simulations at magnitudes around -30 dB. However, at lower magnitudes larger deviations occur for the open in air and short standard. The evaluation shows, that the observed deviations are increased compared to the GGB ISS, suggesting an actual error cancellation between ISS and probe model. However, the occurring deviations are still small enough to achieve a good prediction of the DUT characteristics.
To verify the behaviour and the probe tip model of the CM probe the reverse scenario was used. Therefore, the CM probe was calibrated on a GGB CS-15 ISS and the standards were evaluated. The results are shown in Figs. A.3 and A.4 for the reflection coefficient S_{11} and transmission coefficient S_{21}, respectively. Again, comparing the reflection coefficients S_{11} of Fig. A.3 to the result obtained in Fig. 3.11 shows that the location of the reflection coefficient mainly depends on the utilized calibration substrate and not on the probe tip. However, an offset of the reflect standards between measurement and simulation is clearly visible in the Smith diagram.

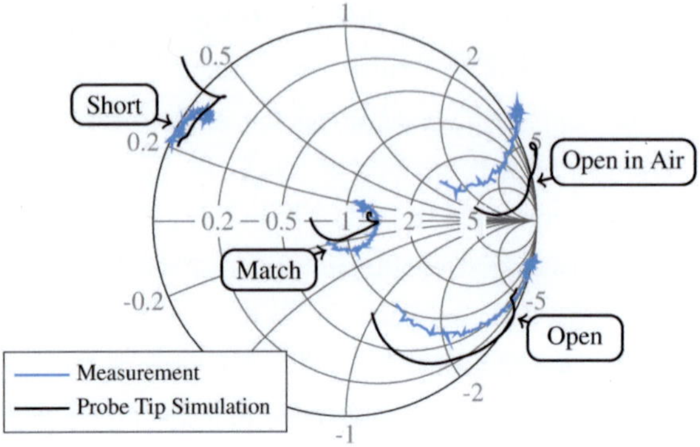

Figure A.3: S_{11} measurement and simulation of GGB CS-15 ISS calibration standards between 200 to 330 GHz contacted with CM probes.

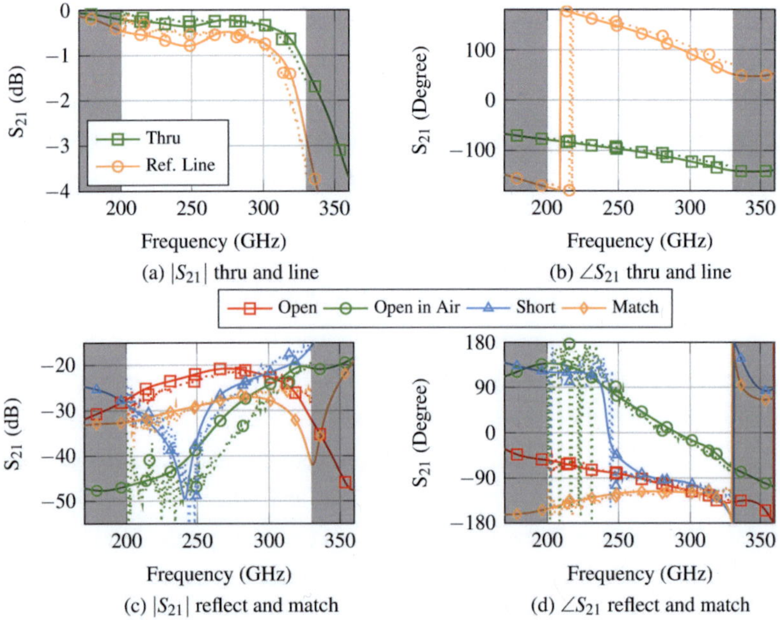

Figure A.4: Measurement (dotted) and simulation (solid) of S_{21} for different GGB CS-15 ISS standards contacted with CM probes.

The transmission and crosstalk behaviour of the calibration standards is depicted in Fig. A.4 and generally shows a good agreement between measurement and simulation results even at low magnitudes. However, it can be observed that the calibrated CS-15 line standards show a strong decrease towards 330 GHz starting around 300 GHz when contacted with CM probes, which does not occur if the GGB probes are used. Comparing the resulting crosstalk it can be seen that using the CM probe on a CS-15 ISS results in an increase with maximum crosstalk magnitudes above −20 dB compared to GGB probes.

A.2 Measurement of different ISS

The last verification step was to first calibrate the used probes with their corresponding ISS. Following that the standards of a different calibration substrate were measured, i.e. CM probes with a CM ISS calibration measured the GGB ISS and vice versa.

Due to the calibration on a different ISS, the thru and line standards can now also be used for evaluation since they are no longer perfectly matched. The measurement and simulation results of the GGB CS-15 standards using the CM probe are shown in Figs. A.5 and A.6, for the input reflection coefficient S_{11} and the transmission coefficient S_{21}, respectively. As can be seen, despite certain deviations the agreement between measurement and simulation is very good. The measured characteristics of the thru and line standard are precisely reproduced with all occurring deviations from the ideal behaviour. Even the crosstalk magnitude measured for the reflect standards, which is below −20 dB, shows a very good agreement between measurement and simulation.

The measured and simulated verification of CM ISS138-357 calibration standards using the GGB probe with a GGB ISS calibration is shown in Figs. A.7 and A.8, for the input reflection coefficient S_{11} and the transmission coefficient S_{21}, respectively.

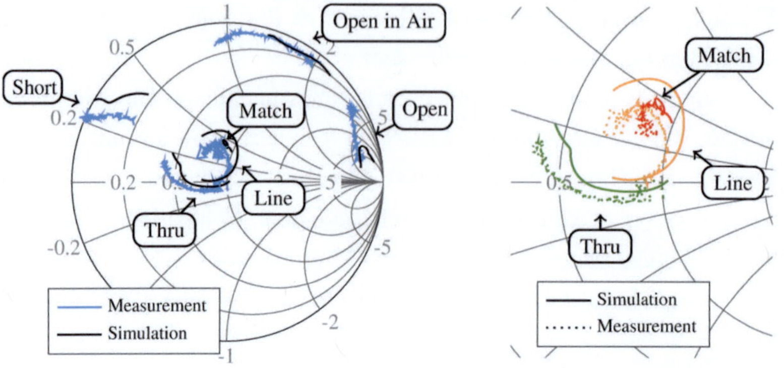

(a) Overview (b) Detail

Figure A.5: S_{11} measurement and simulation of GGB CS-15 ISS between 200 to 330 GHz contacted with CM probes and calibrated on a ISS138-357

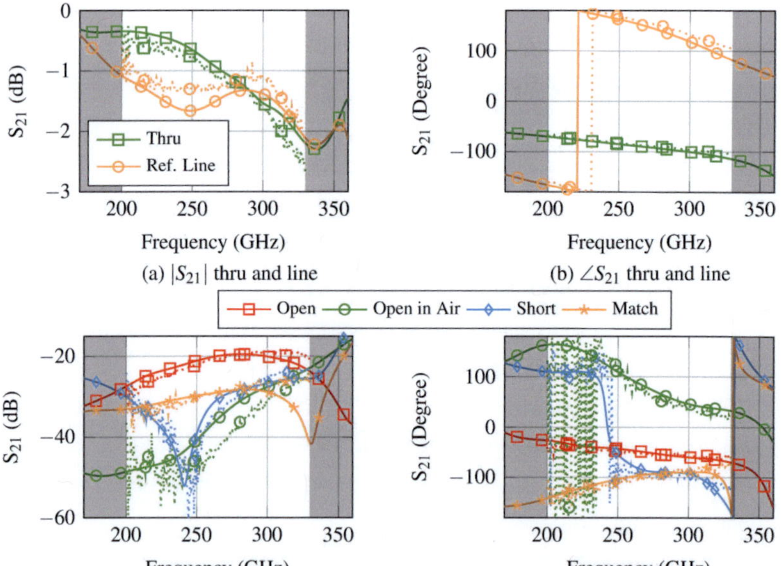

(a) $|S_{21}|$ thru and line (b) $\angle S_{21}$ thru and line

(c) $|S_{21}|$ reflect and match (d) $\angle S_{21}$ reflect and match

Figure A.6: S_{21} measurement (dotted) and simulation (solid) of GGB CS-15 ISS contacted with CM probes calibrated on ISS138-357.

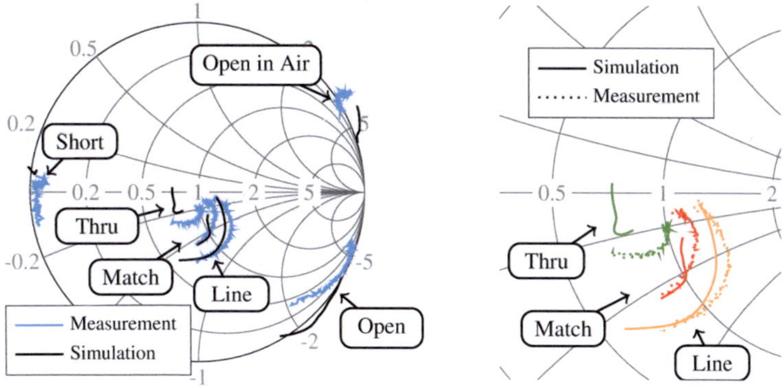

(a) Overview (b) Detail

Figure A.7: S_{11} measurement and simulation of the CM ISS138-357 between 200 to 330 GHz contacted with GGB probes calibrated on a CS-15 ISS

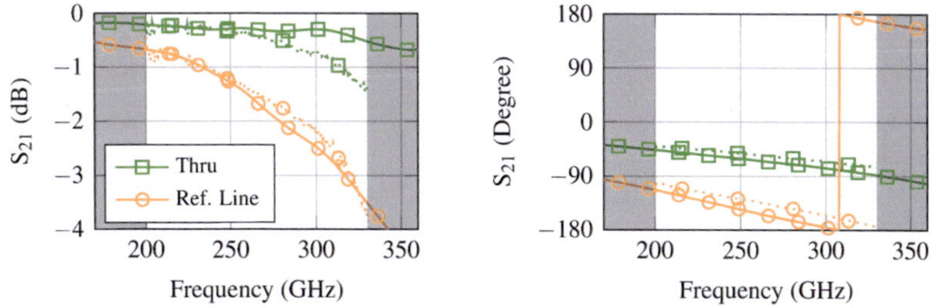

(a) $|S_{21}|$ thru and line (b) $\angle S_{21}$ thru and line

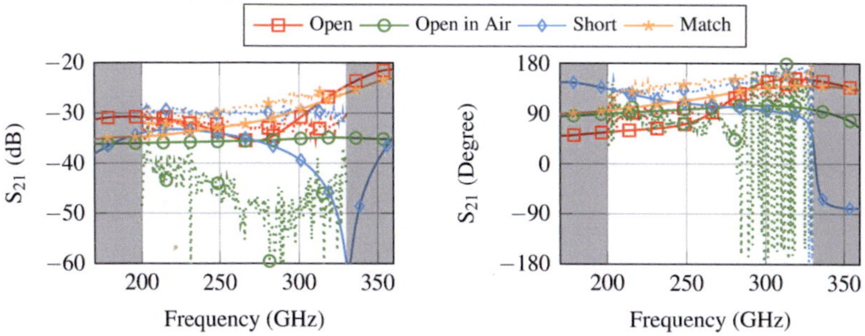

(c) $|S_{21}|$ thru and line (d) $\angle S_{21}$ reflect and match

Figure A.8: S_{21} measurement (dotted) and simulation (solid) of CM ISS138-357 contacted with GGB probes calibrated on CS-15 ISS.

While the simulation and measurement of the line standard are in very good agreement the thru standard shows deviations in S_{21} as well as S_{11}. The crosstalk simulations are in good agreement down to magnitudes of -30 dB while deviating below. Especially the open in air, which was measured with crosstalk down to -50 dB is predicted by the simulation almost frequency independent at -35 dB. However, even though the deviations of the GGB model are larger than for the CM model, the overall agreement between measurement and simulation is still good.

A.3 Conclusion on appendix A

The comparison of the measurement and simulation results of the different investigated scenarios yields the conclusion, that the calibrated behaviour of the calibration standards is mainly dependent on the calibration substrate and not on the used probe. Furthermore, the measurement of a different ISS proved the quality of the probe tip models. However, the results indicate that the CM probe model achieves more accurate results than the GGB probe model. The assumption that the GGB probe has a lower crosstalk compared to the CM probe is confirmed by the simulations and measurements.

B Contact pad measurements with GGB probes

B.1 DGS-pad measurement with GGB probes

The simulation and measurement results of the DGS-pad test structure for different mounting materials when contacted with GGB probes is shown in Fig. B.1. The input reflection coefficient S_{11} is below $-10\,\text{dB}$ up to 300 GHz, independent from the used mounting substrate. While the matching at frequencies around 200 GHz is better compared to the results obtained by CM probes, the CM probe achieves a matching better than $-10\,\text{dB}$ over the full frequency range up to 330 GHz. The behaviour of S_{21} in magnitude and phase is comparable to the CM probe results, with a slightly lower loss in case the GGB probes are used. Analysing the CDR loss however shows that the CM probe results have up to 60 % of loss (Fig. 4.32d), while the GGB probe results stay below 40 % of loss. This indicates that the better matching achieved by the CM probe has no advantage, since it results from radiation of the incident signal.

B.2 TFMS-pad measurement with GGB probes

The simulation and measurement results of the TFMS-pad test structure contacted with GGB probes is shown in Fig. B.2. Analogue to the investigations carried out for the CM probe in section 4.4.3, the test structure was also additionally simulated with vias for substrate mode suppression. Comparing the input reflection coefficient S_{11} in Fig. B.2a indicates that the GGB probe has a matching up to -6 dB while the CM probe stays below -10 dB. As can be seen in Fig. B.2b, the simulated magnitude of S_{21}, if additional vias are used, is around -2 dB and shows a variation of approximately 0.5 dB over the specified frequency range. Without the additional vias several resonances are clearly visible with distortions up to 2 dB. These distortions are also slightly visible in the phase of S_{21}, as shown in Fig. B.2c. However, the CDR loss shown in Fig. B.2 has higher values than that of the CM probe. This indicates that the better input matching of the CM probe works as expected and the incident energy is not radiated as for the DGS-pad.

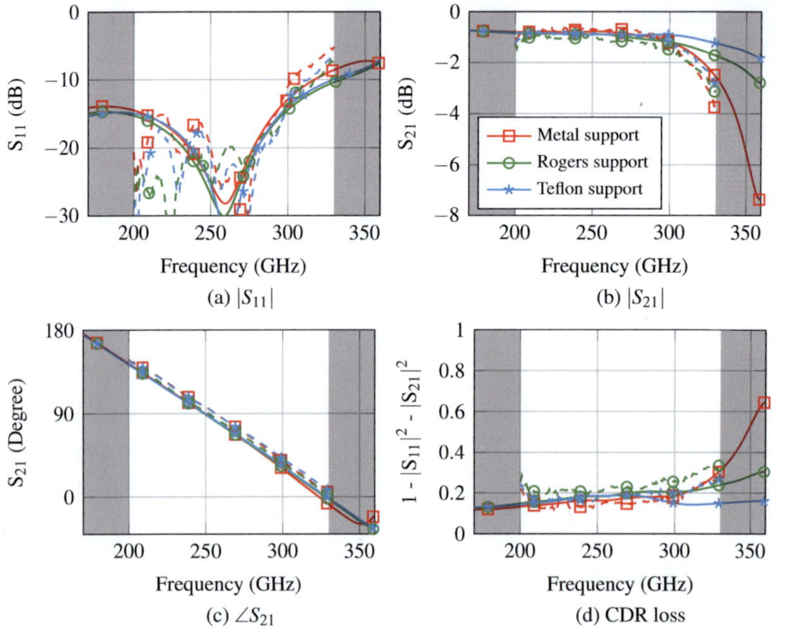

Figure B.1: Measurement (dashed) and simulation (solid) of the test structures for the DGS-pad using GGB probes.

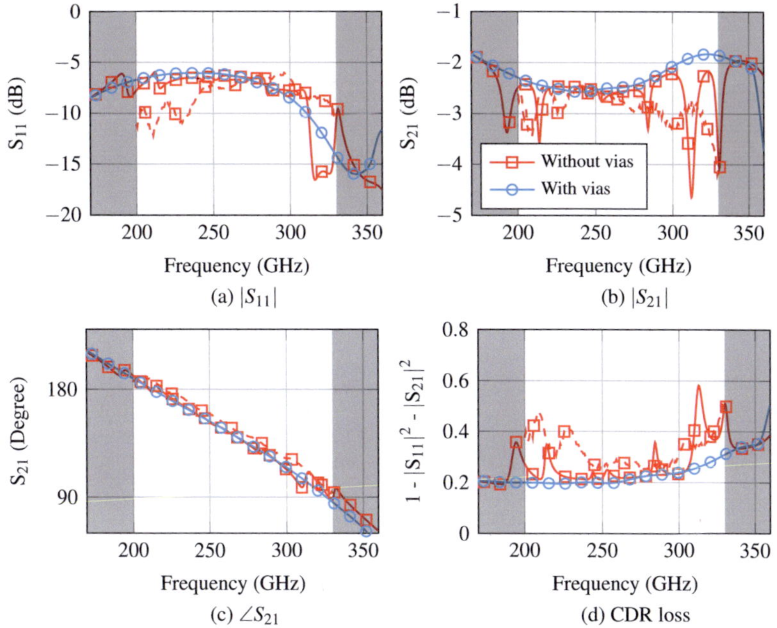

(a) $|S_{11}|$ (b) $|S_{21}|$ (c) $\angle S_{21}$ (d) CDR loss

Figure B.2: Measurement (dashed) and simulation (solid) of the test structures for the TFMS-pad measured with GGB probes.

C DUT circuit topologies and characteristics

C.1 Common phase shifter characteristics

Phase shifters are circuit components, which allow to introduce a controllable phase shift to an incident signal. While so-called analogue phase shifters allow to continuously adjust the phase of the output signal, discrete phase shifters are restricted to discrete phase steps. Usually discrete phase shifters are designed to phase shift the output signal in steps of $360°/2^K$, with K being the number of cascaded phase shifting stages, so-called bits. For example: a 4-bit discrete phase shifter is able to phase-shift the output signal in steps of 22.5°. To control the phase shift, an externally applied voltage V_{ctrl} is used. The key characteristic of any phase shifter is the relative phase shift, which is calculated by referencing the resulting phase of S_{21} for the different control voltages V_{ctrl} to the resulting phase of S_{21} for a fixed control voltage $V_{\text{ctrl,ref}}$.

$$\angle S_{21,\text{rel}} = \angle S_{21} - \angle S_{21,ref} \tag{C.1}$$

Ideally the phase shifter only influences the phase of the applied signal. In practical realization, however, the output signal is distorted in several characteristics. Therefore, to characterize and compare the performance of different phase shifters, several figures of merit (FOM) are used. The mean magnitude of S_{21} is calculated for each frequency over all the full quantity N of different control voltages V_{ctrl}:

$$|\overline{S}_{21}| = \frac{1}{N} \sum_{V_{\mathrm{ctrl}}} |S_{21}| \tag{C.2}$$

For evaluating the broadband characteristics, the RMS phase and amplitude errors are a common FOM. The RMS phase error is calculated by the following steps. A vector of the intended frequency response of the phase shifter is created. In case of a K-bit discrete phase shifter, the intended response is a constant phase shift over frequency with a step size of $360°/2^K$. For analogue phase shifters the intended response can be chosen identical to discrete phase shifters, in case the analogue phase shifter covers a relative phase shift of 0°-360°. Otherwise equal phase steps between 0° and the maximum relative phase shift are chosen for the intended response. The next step is to subtract the different relative phase shift responses from their intended response for all frequencies, resulting in the raw phase error:

$$\angle S_{21,\mathrm{raw,error}} = \angle S_{21,\mathrm{rel}} - \angle S_{21,\mathrm{intended}} \tag{C.3}$$

While the results of (C.3) could in principle be used for the calculation of the RMS phase error, this would lead to wrong results, since the chosen reference state is not the true reference state. To calculate the true reference state the mean raw phase error is subtracted from the raw phase error, which is then used to calculate the corrected phase error:

$$\angle S_{21,\mathrm{corr,error}} = \angle S_{21,\mathrm{raw,error}} - \angle \overline{S}_{21,\mathrm{raw,error}} \tag{C.4}$$

Finally, the RMS phase error is calculated for each frequency with

$$\angle S_{21,\mathrm{RMS,error}} = \sqrt{\frac{1}{N} \sum_{V_{\mathrm{ctrl}}} (\angle S_{21,\mathrm{corr,error}})^2} \tag{C.5}$$

The final step of the calculation of the RMS magnitude error for each frequency is done with

$$|S_{21,\mathrm{RMS,error}}| = \sqrt{\frac{1}{N}\sum_{V_{\mathrm{ctrl}}}(|S_{21}| - |\overline{S}_{21}|)^2} \tag{C.6}$$

Evaluating the RMS errors of a phase shifter gives a detailed insight in the broadband performance and allows to estimate the deviations to the output signal.

C.1.1 Reflective-type phase shifter

The circuit design of a RTPS, shown in Fig. C.1 [Mül+14; Mül+17a], is simple and based on a 90° hybrid coupler, matching networks M and two identical varactor diodes Z_L. Due to the circuit concept, the varactor diodes are isolated from impedance variations at the input and output, similar to the working principle of a balanced amplifier. This allows to separately analyse the influence of the probe to pad interconnect and the performance of the varactor diode. The incoming signal at port 1 is first divided and phase shifted in the 90° hybrid coupler. In the next step the signals travel through the matching networks M and get reflected at the varactors. After travelling through the matching network M and coupler again the reflected waves interfere destructively at the input and constructively at the output of the phase shifter. The destructive interference results in a good input and output matching. The relative phase shift is realized by adjusting the control voltage of the varactors, which have a voltage dependent capacitance behaviour and can be addressed between −1.0 V to 0.5 V. For a given varactor diode, the achievable performance of the RTPS is predominantly dependent on the design of the matching network M located between coupler and varactor diode [Mül+18b].

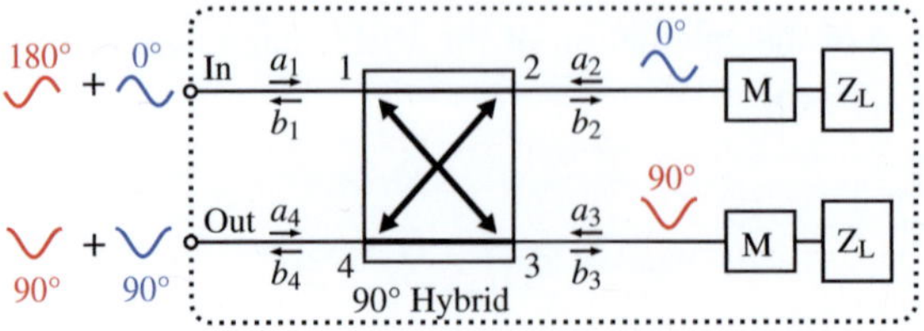

Figure C.1: Basic circuit schematic of a reflective type phase shifter.

C.1.2 Discrete phase shifter

The simplified schematic of the used phase shifter is shown in Fig. C.2a. The circuit consists of a single pole double throw (SPDT) switch, which switches between the two input ports of a distributed transformer (DT) [Pah+15], acting as a 180° coupler.

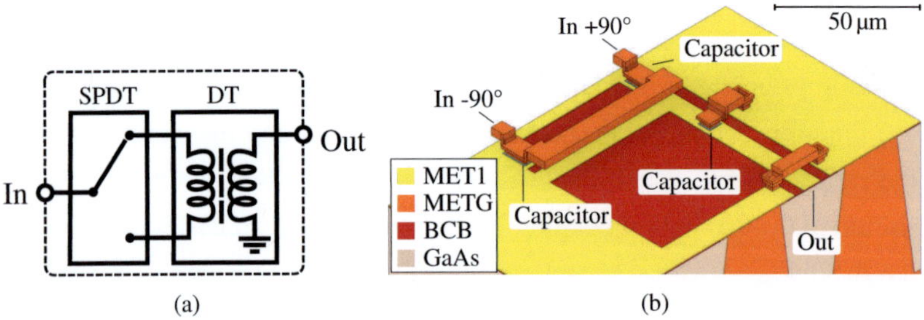

Figure C.2: (a) Simplified schematic of the 180° DPS. (b) EM simulation model of the DT based balun.

C.2 SPDT switch

The simplified schematic of the SPDT switch is shown in Fig. C.3. The basic idea of this concept is to switch an incident signal at the transistor either to ground by biasing the shunt transistors well above pinch-off or to let the signal pass by pinching-off the transistor. For practical realization in the Fraunhofer IAF technology the gate voltage can be adjusted in the range from -1.0 V to 0.5 V. Using two shunt transistors per branch increases the isolation at the cost of a slightly higher insertion loss in the conducting state. A shorted transmission line TL is used to compensate for the off-state capacitance of the shunt transistors [Kal+08]. By the use of a $\lambda/4$ long line the short circuit is converted into an open circuit at the common node of both branches. Therefore, the incident signal at port 1 sees the switched-off branch as an open circuit and the full signal energy travels along the conducting branch. Switching the branches in reverse results in a constant input matching for the common node of port 1. At port 2 the output is well matched if the branch is used as a conducting path but has a high reflection when the branch is isolating. For measurement reasons, the third port of the SPDT switch has to be terminated on-chip with an integrated 50 Ω resistor.

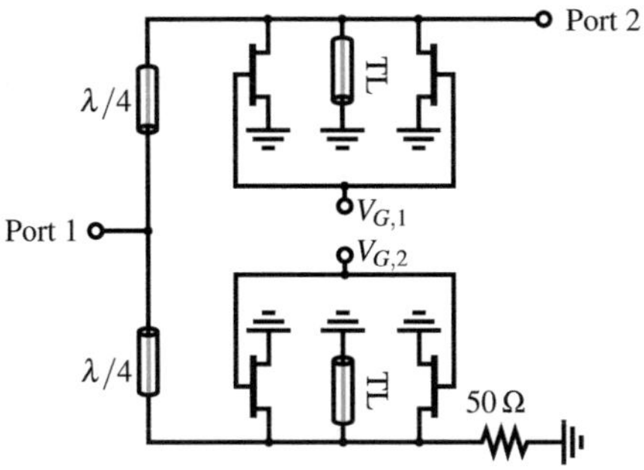

Figure C.3: (a) Schematic view of the SPDT switch.

C.3 Active power divider

The schematic of the APD in Fig. C.4 shows the simplicity of the circuit topology, consisting only of three transistors. At the input the circuit is DC-decoupled by a serial capacitor allowing to cascade the APD to multiple stages with higher power division factors. The first transistor is in common source (CS) configuration and matched to the input port by a simple T-network. At the drain of the CS transistor, the signal is split into two branches which are each connected to a common gate (CG) transistor. The gate widths of the CG transistors are chosen to be equal to the half gate width of the CS transistors. Therefore, the overall current per gate width is constant for the CG and CS transistors, resulting in the same biasing point. The CG transistors are matched to the output port by a second T-network and DC-decoupled with a serial capacitor. The applied drain voltage in the measurements was set to 2.1 V and a drain current of 450 mA/mm was set by adjusting the gate source voltage of the transistors. To investigate the influence of different contact pads three MMICs were realized with the same APD core but different pad structures.

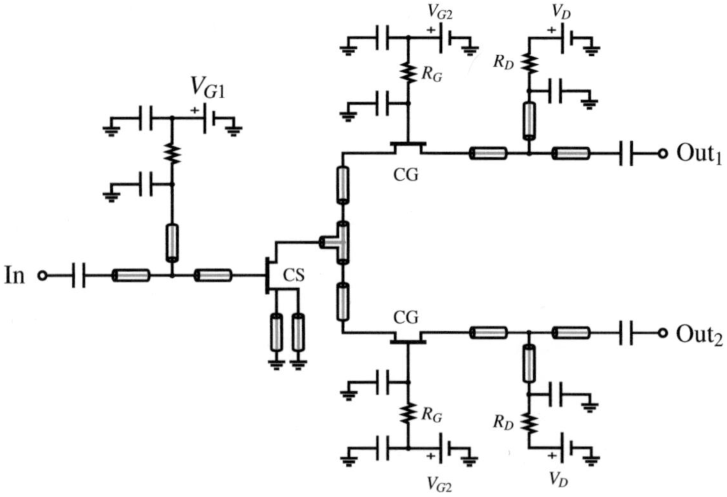

Figure C.4: Schematic of the active power divider core.

C.4 Balanced variable gain amplifier

The amplifier itself is carried out as a balanced amplifier using a broadband Lange coupler for signal splitting and combining, as shown in Fig. C.5. The two amplifying branches are chosen identical, which is essential for a balanced amplifier topology. Due to the identical branches the reflected signals at the individual branches cancel each other out after combined in the coupler, which ideally results in a perfect matching of the overall circuit. Each of the branches consists of three individual amplifying stages, which are realized in a cascode transistor configuration [Die+13]. The detailed schematic of an individual branch is shown in Fig. C.6. In order to adjust the gain of the amplifier the gate-source voltage of the CS and CG transistor are adjusted, leading to a change of transconductance g_m. In order to ensure identical bias points for the CS and CG transistor, the CG gate-source voltage is shifted by half of the drain voltage $V_{gs,cg} = V_{gs,cs} + V_{DD}/2$. The drain voltage of the cascode was chosen to 2 V with a maximum current density of 450 mA/mm.

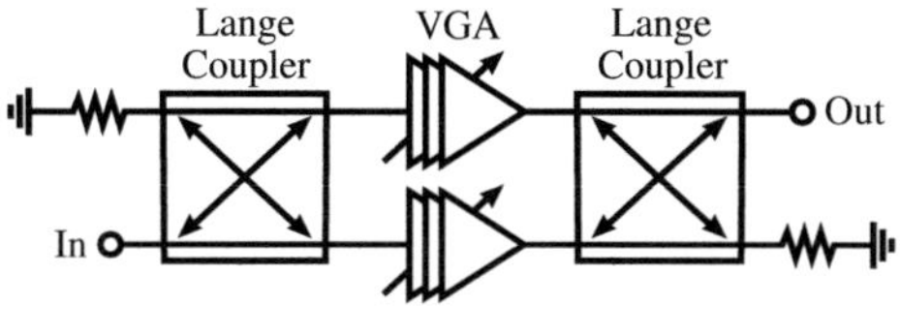

Figure C.5: Simplified schematic of the VGA [Mül+17c].

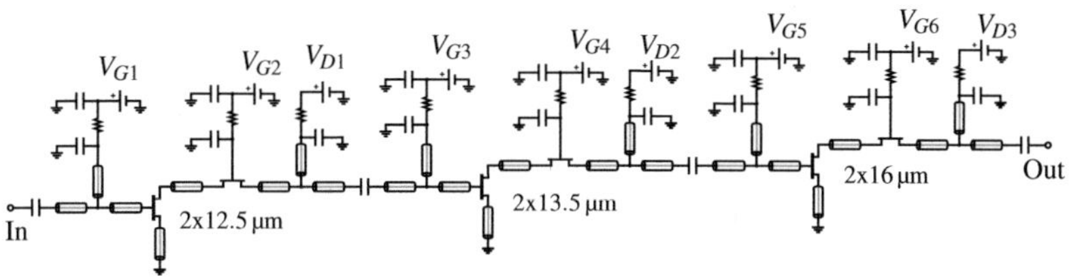

Figure C.6: Detailed schematic of an individual branch of the VGA [Mül+17c].

Bibliography

[Bau+14] M. F. Bauwens, N. Alijabbari, A. W. Lichtenberger, N. S. Barker, and R. M. Weikle. „A 1.1 THz micromachined on-wafer probe“. In: *2014 IEEE MTT-S International Microwave Symposium (IMS2014)*. June 2014, pp. 1–4. DOI: 10.1109/MWSYM.2014.6848607.

[But+91] J. V. Butler, D. K. Rytting, M. F. Iskander, R. D. Pollard, and M. Vanden Bossche. „16-term error model and calibration procedure for on-wafer network analysis measurements“. In: *IEEE Transactions on Microwave Theory and Techniques* 39.12 (Dec. 1991). ISSN: 0018-9480. DOI: 10.1109/22.106567.

[Cam+07] R. Campbell, M. Andrews, L. Samoska, and A. Fung. „Membrane Tip Probes for On-Wafer Measurements in the 220 to 325 GHz Band“. In: *Proc. 18th Int. Symp. Space Terahertz Technology* (2007).

[Cas] Cascade. *Probe Selection Guide*. Cascade Microtech Inc. URL: https://www.cascademicrotech.com/files/Probe-Selection-Guide.pdf (Retrieved 02/01/2018).

[Cas04] Cascade. *The Infinity Probe for On-Wafer Device Characterization and Modeling to 110 GHz*. 2004. URL: https://www.cascademicrotech.com/files/INFINITY_QA.pdf (Retrieved 02/01/2018).

[DJS90] A. Davidson, K. Jones, and E. Strid. „LRM and LRRM Calibrations with Automatic Determination of Load Inductance“. In: *36th ARFTG Conference Digest*. Vol. 18. Nov. 1990, pp. 57–63. DOI: 10.1109/ARFTG.1990.323996.

[Die+13] S. Diebold, D. Müller, D. Schwantuschke, S. Wagner, R. Quay, T. Zwick, and I. Kallfass. „AlGaN/GaN-based variable gain amplifiers for W-band operation“. In: *2013 IEEE MTT-S International Microwave Symposium Digest (MTT)*. Institute of Electrical and Electronics Engineers (IEEE), June 2013. DOI: 10.1109/mwsym.2013.6697340.

[Die+12] S. Diebold, M. Seelmann-Eggebert, H. Gulan, A. Leuther, T. Zwick, and I. Kallfass. „Modelling of transistor feeding structures based on electro-magnetic field simulations“. In: *2012 Workshop on Integrated Nonlinear Microwave and Millimetre-wave Circuits*. Sept. 2012, pp. 1–3. DOI: 10.1109/INMMIC.2012.6331919.

[EH79] G. F. Engen and C. A. Hoer. „Thru-Reflect-Line: An Improved Technique for Calibrating the Dual Six-Port Automatic Network Analyzer“. In: *IEEE Transactions on Microwave Theory and Techniques* 27.12 (Dec. 1979), pp. 987–993. ISSN: 0018-9480. DOI: 10.1109/TMTT.1979.1129778.

[GS15] L. Galatro and M. Spirito. „Analysis of residual errors due to calibration transfer in on-wafer measurements at mm-wave frequencies“. In: *2015 IEEE Bipolar/BiCMOS Circuits and Technology Meeting - BCTM*. Oct. 2015, pp. 141–144. DOI: 10.1109/BCTM.2015.7340569.

[GGB] GGB. *GGB Picoprobe Model 325*. URL: http://www.ggb.com/PdfIndex_files/Model325B.pdf (Retrieved 02/01/2018).

[God92] E. M. Godshalk. „Surface Wave Phenomenon in Wafer Probing Environments“. In: *40th ARFTG Conference Digest*. Vol. 22. Dec. 1992, pp. 10–19. DOI: 10.1109/ARFTG.1992.326994.

[Hie07] M. Hiebel. *Fundamentals of Vector Network Analysis*. Rohde & Schwarz, 2007. ISBN: 9783939837060.

[Jam+14] A. Jam, M. Moallem, J. East, and K. Sarabandi. „A Non-Contact Waveguide Probe for On-Wafer *S*-Parameter Measurements for Submillimeter-Wave to Terahertz Band“. In: *IEEE Transactions on Terahertz Science and Technology* 4.4 (July 2014), pp. 515–522. ISSN: 2156-342X. DOI: 10.1109/TTHZ.2014.2329311.

[Kai05] K. L. Kaiser. *Electromagnetic compatibility handbook*. Boca Raton, Fla. [u.a.]: CRC Press, 2005. ISBN: 0-8493-2087-9.

[Kal+08] I. Kallfass, S. Diebold, H. Massler, S. Koch, M. Seelmann-Eggebert, and A. Leuther. „Multiple-Throw Millimeter-Wave FET Switches for Frequencies from 60 up to 120 GHz“. In: *2008 European Microwave Integrated Circuit Conference*. Oct. 2008, pp. 426–429. DOI: 10.1109/EMICC.2008.4772320.

[KKK17] Mukesh Khandelwal, Binod Kanaujia, and Sachin Kumar. „Defected Ground Structure: Fundamentals, Analysis, and Applications in Modern Wireless Trends“. In: *International Journal of Antennas and Propagation Volume 2017* 2017 (Jan. 2017), pp. 1–22.

[KS71] W. Kruppa and K. F. Sodomsky. „An Explicit Solution for the Scattering Parameters of a Linear Two-Port Measured with an Imperfect Test Set (Correspondence)“. In: *IEEE Transactions on Microwave Theory and Techniques* 19.1 (Jan. 1971), pp. 122–123. ISSN: 0018-9480. DOI: 10.1109/TMTT.1971.1127466.

[Leu+13] A. Leuther, A. Tessmann, M. Dammann, H. Massler, M. Schlechtweg, and O. Ambacher. „35 nm mHEMT Technology for THz and ultra low noise applications“. In: *2013 International Conference on Indium Phosphide and Related Materials (IPRM)*. May 2013, pp. 1–2. DOI: 10.1109/ICIPRM.2013.6562647.

[Leu+07] A. Leuther, A. Tessmann, M. Dammann, C. Schworer, M. Schlechtweg, M. Mikulla, R. Losch, and G. Weimann. „50nm MHEMT Technology for G- and H-Band MMICs“. In: *Indium Phosphide Related Materials, 2007. IPRM '07. IEEE 19th International Conference on*. 2007, pp. 24–27. DOI: 10.1109/ICIPRM.2007.380680.

[MRS01] „Gallium arsenide (GaAs) dielectric constants“. In: *Group IV Elements, IV-IV and III-V Compounds. Part a - Lattice Properties*. Ed. by O. Madelung, U. Rössler, and M. Schulz. Berlin, Heidelberg: Springer Berlin Heidelberg, 2001, pp. 1–11. ISBN: 978-3-540-31355-7. DOI: 10.1007/10551045_113.

[Mar91] R. B. Marks. „A multiline method of network analyzer calibration“. In: *IEEE Transactions on Microwave Theory and Techniques* 39.7 (July 1991), pp. 1205–1215. ISSN: 0018-9480. DOI: 10.1109/22.85388.

[Mül+17a] D. Müller, A. Beck, H. Massler, A. Tessmann, A. Leuther, T. Zwick, and I. Kallfass. „A WR3-band reflective-type phase shifter MMIC with integrated amplifier for error- and loss compensation“. In: *2017 12th European Microwave Integrated Circuits Conference (EuMIC)*. Oct. 2017, pp. 1–4. DOI: 10.23919/EuMIC.2017.8230645.

[Mül+18a] D. Müller, F. Boes, A. Tessmann, A. Leuther, T. Zwick, and I. Kallfass. „Crosstalk Analysis and Correction in On-Wafer Measurements at WR-3 Band Frequencies“. In: *2018 German Microwave Conference*. Institute of Electrical and Electronics Engineers (IEEE), Mar. 2018.

[Mül+17b] D. Müller, M. Fink, J. Eisenbeis, H. Massler, A. Tessmann, A. Leuther, T. Zwick, and I. Kallfass. „A 200 - 300 GHz 1:2 Active power divider MMIC“. In: *2017 42nd International Conference on Infrared, Millimeter, and Terahertz Waves (IRMMW-THz)*. Aug. 2017, pp. 1–2. DOI: 10.1109/IRMMW-THz.2017.8067040.

[Mül+17c] D. Müller, M. Fink, C. V. Vangerow, A. Tessmann, A. Leuther, T. Zwick, and I. Kallfass. „A 220 - 260 GHz medium power variable gain amplifier MMIC with low phase variation“. In: *2017 IEEE Asia Pacific Microwave Conference (APMC)*. Nov. 2017, pp. 29–32. DOI: 10.1109/APMC.2017.8251369.

[Mül+18b] D. Müller, A. Haag, A. Bhutani, A. Tessmann, A. Leuther, T. Zwick, and I. Kallfass. „Bandwidth Optimization Method for Reflective-Type Phase Shifters“. In: *IEEE Transactions on Microwave Theory and Techniques* 66.4 (Apr. 2018), 1754–1763. ISSN: 0018-9480. DOI: 10.1109/TMTT.2017.2779156.

[Mül+14] D. Müller, S. Reiss, H. Massler, A. Tessmann, A. Leuther, T. Zwick, and I. Kallfass. „A h-band reflective-type phase shifter MMIC for ISM-Band applications“. In: *2014 IEEE MTT-S International Microwave Symposium (IMS2014)*. Institute of Electrical and Electronics Engineers (IEEE), June 2014. DOI: 10.1109/mwsym.2014.6848423.

[Mül+17d] D. Müller, J. Schäfer, D. Geenen, H. Massler, A. Tessmann, A. Leuther, T. Zwick, and I. Kallfass. „Electromagnetic field simulation of MMICs including RF probe tips“. In: *2017 47th European Microwave Conference (EuMC)*. Oct. 2017, 900–903. DOI: 10.23919/EuMC.2017.8230990.

[Mül+18c] D. Müller, J. Schäfer, H. Massler, M. Ohlrogge, T. Zwick, and I. Kallfass. „Impact of Ground Via Placement in On-Wafer Contact Pad Design up to 325 GHz“. In: *IEEE Transactions on Components, Packaging and Manufacturing Technology* PP.99 (2018), pp. 1–11. ISSN: 2156-3950. DOI: 10.1109/TCPMT.2018.2811482.

[OML] OML. Inc. *WR3 Frequency Extension Modules*. URL: https://www.omlinc.com/images/pdf/VxxVNA2/OML-V03VNA2_Datasheet.pdf (Retrieved 02/01/2018).

[Pah+15] P. Pahl, S. Wagner, H. Massler, S. Diebold, A. Leuther, I. Kallfass, and T. Zwick. „A 50 to 146 GHz Power Amplifier Based on Magnetic Transformers and Distributed Gain Cells“. In: *IEEE Microwave and Wireless Components Letters* 25.9 (2015).

[Poz12] D. M. Pozar. *Microwave engineering*. 4. ed. Hoboken, NJ: Wiley, 2012. ISBN: 978-0-470-63155-3.

[Saf05] A. M. E. Safwat. „Study of microstrip mode in RF on-wafer probes“. In: *Microwave and Optical Technology Letters* 45.4 (2005), pp. 324–328. ISSN: 1098-2760. DOI: 10.1002/mop.20812.

[SH02] A. M. E. Safwat and L. Hayden. „Sensitivity analysis of calibration standards for fixed probe spacing on-wafer calibration techniques [vector network analyzers]“. In: *2002 IEEE MTT-S International Microwave Symposium Digest (Cat.*

No.02CH37278). Vol. 3. June 2002, 2257–2260 vol.3. DOI: 10.1109/MWSYM.2002.1012323.

[Sch+11] F. J. Schmückle, R. Doerner, G. N. Phung, W. Heinrich, D. Williams, and U. Arz. „Radiation, multimode propagation, and substrate modes in W-band CPW calibrations“. In: *2011 41st European Microwave Conference*. Oct. 2011, pp. 297–300.

[See+15] M. Seelmann-Eggebert, M. Ohlrogge, R. Weber, D. Peschel, H. Maßler, M. Riessle, A. Tessmann, A. Leuther, M. Schlechtweg, and O. Ambacher. „On the Accurate Measurement and Calibration of S-Parameters for Millimeter Wavelengths and Beyond“. In: *IEEE Transactions on Microwave Theory and Techniques* 63.7 (July 2015), 2335–2342. ISSN: 0018-9480. DOI: 10.1109/TMTT.2015.2436919.

[Sim01] R. N. Simons. *Coplanar waveguide circuits, components, and systems*. Wiley series in microwave and optical engineering. New York, NY: Wiley-Interscience, 2001. ISBN: 9780471161219.

[SGA13] M. Spirito, G. Gentile, and A. Akhnoukh. „Multimode analysis of transmission lines and substrates for (sub)mm-wave calibration“. In: *82nd ARFTG Microwave Measurement Conference*. Nov. 2013, pp. 1–6. DOI: 10.1109/ARFTG-2.2013.6737356.

[Tec17] CST - Computer Simulation Technology. 2017. URL: https://www.cst.com/.

[Tes+08] A. Tessmann, A. Leuther, H. Massler, M. Kuri, and R. Loesch. „A Metamorphic 220-320 GHz HEMT Amplifier MMIC“. In: *2008 IEEE Compound Semiconductor Integrated Circuits Symposium*. Oct. 2008, pp. 1–4. DOI: 10.1109/CSICS.2008.12.

[VDI] VDI. *Vector Network Analyzer Extenders.* URL: https : / / vadiodes . com / en / products / vector - network - analyzer - extension-modules (Retrieved 02/01/2018).

[Wil+14a] D. F. Williams, F. J. Schmückle, R. Doerner, G. N. Phung, U. Arz, and W. Heinrich. „Crosstalk Corrections for Coplanar-Waveguide Scattering-Parameter Calibrations“. In: *IEEE Transactions on Microwave Theory and Techniques* 62.8 (Aug. 2014), pp. 1748–1761. ISSN: 0018-9480. DOI: 10.1109/ TMTT.2014.2331623.

[Wil+14b] D. F. Williams, F. J. Schmückle, R. Doerner, G. N. Phung, U. Arz, and W. Heinrich. „Crosstalk Corrections for Coplanar-Waveguide Scattering-Parameter Calibrations“. In: *IEEE Transactions on Microwave Theory and Techniques* 62.8 (Aug. 2014). ISSN: 0018-9480. DOI: 10.1109/TMTT.2014. 2331623.

[Yee66] K. Yee. „Numerical solution of initial boundary value problems involving maxwell's equations in isotropic media“. In: *IEEE Transactions on Antennas and Propagation* 14.3 (May 1966), pp. 302–307. ISSN: 0018-926X. DOI: 10.1109/TAP. 1966.1138693.

[ZG11] T. Zelder and B. Geck. „Contactless Scattering Parameter Measurements“. In: *IEEE Microwave and Wireless Components Letters* 21.9 (Sept. 2011), pp. 504–506. ISSN: 1531-1309. DOI: 10.1109/LMWC.2011.2162619.

[ZL08] K. Zhang and D. Li. *Electromagnetic theory for microwaves and optoelectronics.* 2. ed. Berlin: Springer, 2008. ISBN: 978-3-540-74295-1.

Forschungsberichte aus dem Institut für Höchstfrequenztechnik und Elektronik (IHE) der Universität Karlsruhe (TH)
(ISSN 0942-2935)

Herausgeber: Prof. Dr.-Ing. Dr. h.c. Dr.-Ing. E.h. mult. Werner Wiesbeck

Die Bände 1 (1992) bis 55 (2008) der Schriftenreihe können über das Institut Hochfrequenztechnik und Elektronik bestellt werden (https://www.ihe.kit.edu).

Fortführung als:

Karlsruher Forschungsberichte aus dem Institut für Hochfrequenztechnik und Elektronik
(ISSN 1868-4696)

Herausgeber: Prof. Dr.-Ing. Thomas Zwick

Band 55 Sandra Knörzer
Funkkanalmodellierung für OFDM-Kommunikationssysteme bei Hochgeschwindigkeitszügen (2009)
ISBN 978-3-86644-361-7

Band 56 Thomas Fügen
Richtungsaufgelöste Kanalmodellierung und Systemstudien für Mehrantennensysteme in urbanen Gebieten (2009)
ISBN 978-3-86644-420-1

Band 57 Elena Pancera
Strategies for Time Domain Characterization of UWB Components and Systems (2009)
ISBN 978-3-86644-417-1

Band 58 Jens Timmermann
Systemanalyse und Optimierung der Ultrabreitband-Übertragung (2010)
ISBN 978-3-86644-460-7

Die Bände sind unter www.ksp.kit.edu als PDF frei verfügbar oder als Druckausgabe bestellbar.

Band 59 Juan Pontes
Analysis and Design of Multiple Element Antennas for Urban Communication (2010)
ISBN 978-3-86644-513-0

Band 60 Andreas Lambrecht
True-Time-Delay Beamforming für ultrabreitbandige Systeme hoher Leistung (2010)
ISBN 978-3-86644-522-2

Band 61 Grzegorz Adamiuk
Methoden zur Realisierung von dual-orthogonal, linear polarisierten Antennen für die UWB-Technik (2010)
ISBN 978-3-86644-573-4

Band 62 Jutta Kühn
AlGaN/GaN-HEMT Power Amplifiers with Optimized Power-Added Efficiency for X-Band Applications (2011)
ISBN 978-3-86644-615-1

Band 63 Małgorzata Janson
Hybride Funkkanalmodellierung für ultrabreitbandige MIMO-Systeme (2011)
ISBN 978-3-86644-639-7

Band 64 Mario Pauli
Dekontaminierung verseuchter Böden durch Mikrowellenheizung (2011)
ISBN 978-3-86644-696-0

Band 65 Thorsten Kayser
Feldtheoretische Modellierung der Materialprozessierung mit Mikrowellen im Durchlaufbetrieb (2011)
ISBN 978-3-86644-719-6

Band 66 Christian Andreas Sturm
Gemeinsame Realisierung von Radar-Sensorik und Funkkommunikation mit OFDM-Signalen (2012)
ISBN 978-3-86644-879-7

Band 67 Huaming Wu
Motion Compensation for Near-Range Synthetic Aperture Radar Applications (2012)
ISBN 978-3-86644-906-0

Band 68 Friederike Brendel
Millimeter-Wave Radio-over-Fiber Links based on Mode-Locked Laser Diodes (2013)
ISBN 978-3-86644-986-2

Band 69 Lars Reichardt
Methodik für den Entwurf von kapazitätsoptimierten Mehrantennensystemen am Fahrzeug (2013)
ISBN 978-3-7315-0047-6

Band 70 Stefan Beer
Methoden und Techniken zur Integration von 122 GHz Antennen in miniaturisierte Radarsensoren (2013)
ISBN 978-3-7315-0051-3

Band 71 Łukasz Zwirełło
Realization Limits of Impulse-Radio UWB Indoor Localization Systems (2013)
ISBN 978-3-7315-0114-5

Band 72 Xuyang Li
Body Matched Antennas for Microwave Medical Applications (2014)
ISBN 978-3-7315-0147-3

Band 73 Sebastian Diebold
Transistor- und Leitungsmodellierung zum Entwurf von monolithisch integrierten Leistungsverstärkern für den hohen Millimeterwellen-Frequenzbereich (2014)
ISBN 978-3-7315-0161-9

Band 74 Christian Rusch
Integrierte, planare Leckwellenantennen für 3D-Millimeterwellen-Radarsysteme basierend auf dem holografischen Prinzip (2014)
ISBN 978-3-7315-0234-0

Band 75 Marlene Harter
Dreidimensional bildgebendes Radarsystem mit digitaler Strahlformung für industrielle Anwendungen (2014)
ISBN 978-3-7315-0249-4

Band 76 Michael A. Baldauf
Abhängigkeit der Exposition von der Zellgröße beim Mobilfunk unter Gewährleistung der Versorgung (2015)
ISBN 978-3-7315-0308-8

Band 77 Alicja Ossowska
Highly Resolved Synthetic Aperture Radar with Beam Steering (2015)
ISBN 978-3-7315-0315-6

Band 78 Małgorzata Dominika Brzeska
RF Modelling and Characterization of Tyre Pressure Sensors and Vehicle Access Systems (2015)
ISBN 978-3-7315-0348-4

Band 79 Ulrich Lewark
Aktive Frequenzvervielfacher zur Signalerzeugung im Millimeter- und Submillimeterwellen Frequenzbereich (2015)
ISBN 978-3-7315-0354-5

Band 80 Kai-Philipp Walter Pahl
Distributed Transformers for Broadband Monolithic Millimeter-Wave Integrated Power Amplifiers (2015)
ISBN 978-3-7315-0409-2

Band 81 Serdal Ayhan
Hochgenaue radarbasierte Abstandsmessung mit geführter Wellenausbreitung (2016)
ISBN 978-3-7315-0433-7

Band 82 Yoke Leen Sit
MIMO OFDM Radar-Communication System with Mutual Interference Cancellation (2017)
ISBN 978-3-7315-0599-0

Band 83 Steffen Scherr
FMCW-Radarsignalverarbeitung zur Entfernungsmessung mit hoher Genauigkeit (2017)
ISBN 978-3-7315-0607-2

Band 84 Tom Schipper
Modellbasierte Analyse des Interferenzverhaltens von Kfz-Radaren (2017)
ISBN 978-3-7315-0639-3

Band 85 Malyhe Jalilvand
Application-Specific Broadband Antennas for Microwave Medical Imaging (2017)
ISBN 978-3-7315-0664-5

Band 86 Benjamin Göttel
Millimeterwellen On-Chip Antennensysteme für die Integration in SoC Applikationen (2017)
ISBN 978-3-7315-0667-6

Band 87 Christian Arnold
Im Orbit einstellbare Ausgangsfilter und -multiplexer (2017)
ISBN 978-3-7315-0722-2

Band 88 Tobias Mahler
Synthese kapazitätsoptimierter Antennensysteme mit messtechnischer Verifikation (2018)
ISBN 978-3-7315-0737-6

Band 89 Daniel Müller
RF Probe-Induced On-Wafer Measurement Errors in the Millimeter-Wave Frequency Range (2018)
ISBN 978-3-7315-0822-9